GT3 AT 25

Perhaps the greatest revelation of the water-cooled era, the 911 GT3 is now a quarter century old. Total 911 assesses the legacy of the first six generations

Written by **Kyle Fortune** & **Lee Sibley** Photography by **Ali Cusick**

Mention 'GT3' and Porsche's now-legendary moniker conjures a host of vivid adjectives: Loud. Unrestrained. Pure. Mechanical. Fast.

Porsche's GT3 is already considered an icon – an exemplary feat given it's only just turning 25 years old. Launched just before the turn of the millennium, Porsche's new 911 model line had already positively asserted itself by breaking the Nürburgring lap record for production vehicles with a time of seven minutes and 56 seconds, thereby firing its way straight into the hearts of admiring enthusiasts.

Built to homologate Porsche's FIA race cars, the GT3 was originally built for the UK and mainland Europe only, yet the line-up has since flourished into a worldwide motoring phenomenon, each new model a highlight within its generation of 911.

Total 911 has gathered the first six generations of GT3 for a special test, as we relive 25 years of a special sports car perennially at the peak of its class. Beginning, of course, with the 996 of 1999... ➲

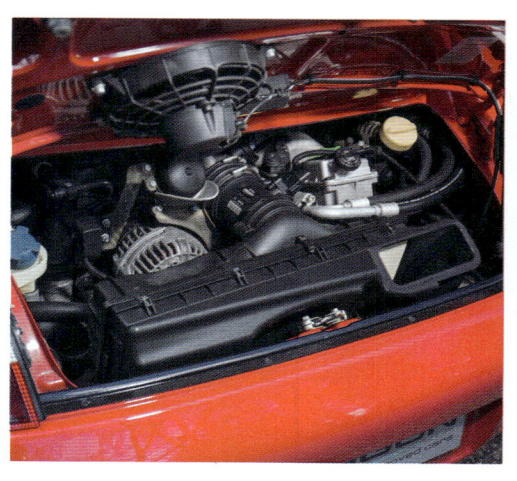

THE **996** YEARS

Simple, compact, powerful and light. Actually, we'll skip the last one, as one 996 factoid that's oft repeated is that the original 996 GT3 was actually heavier than its standard Carrera relation. Heresy in sporting 911 circles, but what the GT3 added in bulk it made up for in muscle and specification, this genesis GT3 creating the direction that the five cars that join it here today would follow.

Central to the GT3 is the Motorsport department's fitment of the M96/76 dry-sumped flat six engine. More commonly referred to as 'the Mezger' after engineer Hans Mezger, no other engine – with the possible exception of Paul Rosche and the McLaren F1's V12 – is so closely associated with an engineer. It's pure motorsport, the 3.6-litre flat six featuring lighter internals and special low-friction surfaces, a crankcase derived from the Le Mans-winning GT1 and the four-valve heads related to those of the 959. As pedigrees go… to that Porsche fitted the G96/90 gearbox, its additional weight, as well as the extra cooling the engine required, being instrumental in the GT3's slight weight gain.

Rated at 360hp at 7,200rpm, with its redline at 7,800rpm, the numbers say it'll reach 62mph in 4.8 seconds before heading to a 188mph top speed. Visually it's uncomplicated: narrow bodied, the rear wing adjustable, the interior stark without being austere. It sits more purposefully, the 996 Gen1 GT3 riding 30mm lower than a standard Carrera on 18-inch split-rim alloy wheels, with 225/40/ZR18 front and 285/30/ZR18 rear tyres.

The suspension specification reads like that of the racer it homologates, so there are adjustable anti-roll bars and a spring system compatible with racing springs, an extended range of axle geometry adjustment and reinforced front pivots, mounts and wheel bearings to cope with the additional stress slick tyres would add if fitted.

Given its unashamedly race-focused specification, the first GT3 presents few compromises. It's stiff, but not at the expense of ride comfort, the suspension supple, the wheel and body control beautifully resolved on the road. It feels small – and not just in this company – the 996's narrow body, upright screen and slim pillars signalling its vintage, such as it is given the GT3 was launched in 1999. As do the lack of electronics – there's not a Sport, PASM, traction or stability control button in sight. It was as pure as driving gets, and that was pretty much the point.

We've driven a few, but this example is among the very best we've been in. Firing up that engine brings the familiar, somewhat mechanical chatter from behind, that very much in keeping with its motorsport DNA. There's not a huge amount of ➲

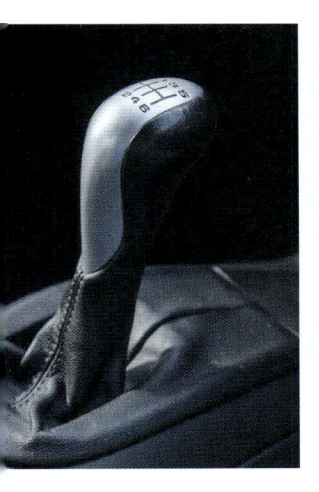

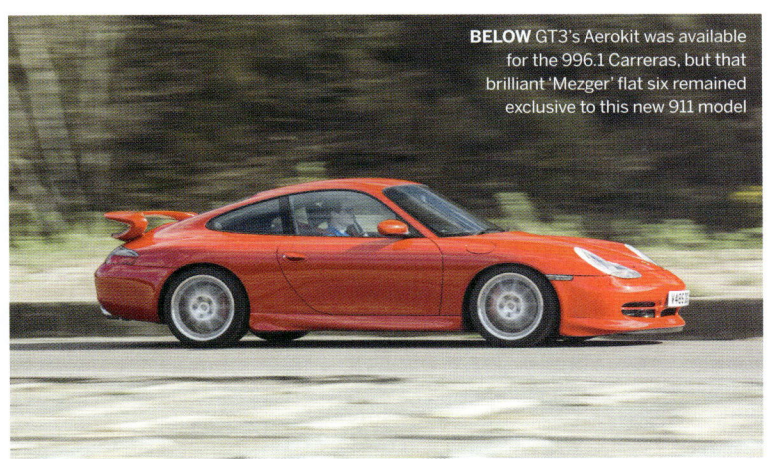

BELOW GT3's Aerokit was available for the 996.1 Carreras, but that brilliant 'Mezger' flat six remained exclusive to this new 911 model

"More commonly referred to as 'the Mezger' after engineer Hans Mezger, no other engine is so closely associated with an engineer"

urgency below 5,000rpm; it's quick enough, but it's not startlingly removed from a standard 996 Carrera. That changes. Get the rev-counter needle higher and the 996.1 car flies, the engine's character taking a far more urgent, harder purpose, bringing with it scintillating acceleration and a soundtrack to match.

The gearshift is light and precise, its throw surprisingly long, the clutch weighty enough to be tiresome for manoeuvring but perfectly judged on the move. It's the steering as much as anything else that wows. It's so detailed, the clarity of the information rich, the response requiring a bit of patience. The Gen1 is a car that drives like an old-school 911, where weight-shift is your friend to tuck the nose in. Despite its lack of electronic assistance that's not something to fear: it's exploitable, wonderful, even, at its very core a driver's car that demands and rewards in ever-greater measure as you drive it harder.

If we'd only driven the Gen1 car today we wouldn't have felt in any way deprived, but there's work to be done. We can feel your collective hearts bleeding. We love how the Gen2 car looks. That simple spoiler, those headlights, the neater skirts. Its shape describes our perfect 996. The technical specification isn't so different, though in the four years that separate the red and the blue car, Porsche Motorsport learned some new tricks.

Key among the changes is the drive-by-wire system, it helping increase the engine's output to 381hp. It's not the bump in power that's obvious, however, but the increased torque, which although only increases by 15Nm to 385Nm is available across more of the engine's range. Low-rev immediacy is greater, yet it's still a screamer up to its 400rpm-higher redline. There's more focus in the suspension, as well as more opportunity to adjust it, while at its nose there's an extra inch of rubber on the road, thanks to the front wheels gaining half an inch each. The rear wheels too grow to 11 inches, and there's GT3-developed brakes, with those on the nose gaining an extra pair of pistons for a count of six.

Still delicate and pure, the Gen2's changes make for a car that feels significantly faster, and you're still on your own. The limits are higher, the detail still there, but the traits of the Gen2 are accessed at a slightly higher level. The nose still needs some work to get tucked in, but the rear feels a bit more settled at ordinary speeds, and the brakes are stronger. Better, undoubtedly, but that doesn't see us wanting that Gen1 car any less.

BELOW The 996.2 has the privilege of being the GT3 with (on paper) the closest spec and performance stats to its bigger RS brother

THE 997 YEARS

Moving from 996 to 997 defines a period of rapid advancement at Porsche. Never before had its 911 model changed so much in such a short space of time. In 2006 the dashboard in the 997 introduced modernity, the 996's relative simplicity and ease making way for the complexity of a screen, and more buttons than a Pearly Queen's hat.

That dates both the 996 and 997 inside, the 996 by its pre-era connectivity and simplicity, the 997-era cars spanning early modernity, and it shows. As does the standard air-bagged wheel, which looks heavy-handed in comparison to the neat, slim-spoked yet still air bag-equipped steering wheel of the 996. Details are, frankly, irrelevant, as you buy a GT3 to drive, not to press buttons. The 997 introduced a couple of significant ones though, with Porsche Active Suspension Management (PASM) added to the chassis specification, and electronic traction control, which Porsche claims was adapted from the Carrera GT. That traction control incorporates traction slip and drag-torque control among its electronic trickery, though hold the button and you're on your own.

> "There's more grip, more aero, more power… more everything with the 997 GT3"

Like its predecessors there's a mechanical limited-slip differential with 28 per cent locking under load and 40 per cent on overrun, while the engine remains the 3.6-litre. Revisions to the Mezger unit see its maximum revs rise a further 200rpm to 8,400rpm, that achieved by improving the intake, with the throttle valve growing considerably from 76 to 82mm. Reduced back pressure also helps, Porsche quoting an output of 415hp, some 115.3hp per litre, which is an incredible specific output. 62 miles per hour is now possible in 4.3 seconds, aided by a lower ratio second gear – sixth also being lower – the shift itself shorter and a shift light added to the rev-counter in case the 3.6-litre flat six screaming behind you isn't enough of a reminder to grab another ratio.

Where the 996s are relatively subtle visually, there's a pugnacious, overtly motorsport look to the 997.1 GT3. Aero plays a far more decisive role in its styling, the jutting black splitter under the many-apertured bumper itself topped by a vent to manage cooling airflow over the front of the car. There are punctured vents in the rear, the wing large above the engine – and in profile echoing the shape of the 996.1. Wheels go up to 19-inches, with 235/35/

Comfort v Clubsport

Cage, seats, harnesses and a fire extinguisher. Your call, but for many the Clubsport package is a must-have option with the GT3. We're not one of them; yes, the GT3's appeal is unquestionably its track ability, but for the majority of the time you'll likely be driving it on the road. A cage adds inconvenience if you're ever using the rear deck for luggage, while they're all a bit noisier on the road with some pipework behind the seats. Lightweight bucket seats are the only must-have option in our opinion, not least because we find the greater adjustability of the regular seats only adds the opportunity for discomfort, but the key reason is a bucket increases the connection. That's us; you might differ. There is too much nonsense talked about whether Clubsport is greater than a Comfort and vice versa. Buy the GT3 you desire, and don't feel like you need to justify it by questioning those who choose differently. They're all exceptional, Clubsport or not.

ABOVE It was all change as the GT3 moved from 997 to 991, the latter gaining an active rear axle, electrically-assisted steering, and PDK-only transmission
BELOW Centre-locking wheels featured for the first time on the 997.2

ZR19 front and 305/30/ZR19 tyres. There's more grip, more aero, more power... more everything with the 997.1 GT3.

After the delicacy of the 996s, the 997's 'greater everything' could be its undoing. It isn't, the 997's neatest trick being that it retains the incredible control and feel of its predecessor, but elevates the performance to another level. The numbers associated with it are, arguably, incidental. It's the feel that the 997 offers that defines the drive, dominating the proceedings and involving the driver like nothing else. If there's one element that underlines the GT3 it's the retention of a manual transmission, the six-speeder an anachronism in a paddle-shifted world, yet it's central, and core to its appeal.

You have to drive the GT3; there's no opportunity to switch off, that appealing as engagement rather than being tiresome. The clutch is heavy, the shift it allows being quick and precise, and the engine – damn, that engine. There's greater low-rev urgency, the still rev-happy 3.6-litre being tractable even if you're not chasing that redline. Just try doing that though: the GT3's unit goads you into wringing it out, the reward when you do so being little short of mechanical nirvana.

Sensational as that engine is, it's not dominant. The chassis is its measure, creating a perfect symbiosis where the engine's forces can be exploited, the result being incredible, pure speed, and control that's in the other-worldly sphere. You don't need anything faster, but just three years later Porsche would introduce its Gen2 model.

More power, now 435hp, from an engine enlarged to 3.8 litres, the increased performance coming with the bonus of improved economy and emissions. There are more revs, too, the maximum now 8,500rpm, that a given, while the creep of electronic control is raised with the addition of Porsche Stability Management (PSM), incorporating Stability Control and Traction Control. It's switchable, partially or all off, though even 'on' it's judged so finely that it can genuinely be considered an aid rather than a hindrance.

You'll be doing well to require them on the road, in the dry at least, the 997's grip and traction phenomenal, the lighter, now centre-lock wheels increasing the wheel control, the suspension managing the fine line between control and comfort exceptionally well. If the Gen1 car is mesmerising, then its replacement elevates that status even further, the fine balance, incredible control and the feel and weighting all marking it out as one of the greatest, puristic driver's cars you can buy. Yes, there are RS models above, but we defy anyone to climb out of a 997 GT3 – either generation – and not want to immediately get back in and drive it some more.

18 | Ultimate 911 GT3 Collection

THE 991 YEARS

The shift was seismic. We remember sitting in the press conference at launch with Andreas Preuninger justifying the loss of the manual transmission. "It's faster," he said, not looking entirely convinced. The 991 was always going to be a different GT3, but to what degree? A bigger car physically, the 991 adopted the Carrera 4's wide body, lost the Mezger engine for a 3.8-litre derivation of the Carrera S's unit and featured electrically assisted power steering over the hydraulic set-up of its predecessors. The loss of manual transmission was its most significant change, one Porsche would rectify with the Gen2.

Does that leave the 991.1 GT3 as something to be overlooked? Not one bit, it bringing more breadth to the GT3's repertoire yet retaining most of its core appeal. Power is now 475hp from an engine that might not be the Mezger in actuality, but apes its enthusiasm for revs and keenness of response with a 9,000rpm redline. Agility remains sensational, too, though here the 991 shrinks itself by throwing a technical arsenal at the drive, including rear-wheel steering in the mix, Preuninger describing the GT3 at launch as: "Reloaded, updated, but still refreshingly different."

He wasn't wrong. The ease by which the 991.1 GT3 can carry its speed is incredible, it retaining all the elementary traits that define the GT3, those being poise, balance, speed, sensational brakes and an engine that's spine-tingling in its response, feeling directly connected to your synapses. It's easier though, the digit-flicking, gear-shifting capabilities that the PDK brought adding another dimension to the drive as much as they rob the old-school purists of one of the key differentiators of the cars that preceded it. That's both a compliment and a ➲

> "The Touring needs all of your attention, all of your time, the PDK car allowing you some respite should you require it"

BELOW The 991.1 changed the game for the GT3, offering a car which had day-to-day civility alongside blistering performance and balance when called upon

GT3 at 25 | 19

BELOW Touring Package gives a traditional 911 silhouette to the GT3, the usual fixed wing replaced with a small Gurney flap on top of an active aero aid

complaint, the GT3 now a more usable car, whether that's daily or hardcore track use. Grip levels are sensational, traction mighty, yet there's delicacy and feel, even from the steering, which is quite an achievement with an electronically assisted power-steering system. The PDK allowed the fitment of an electronically controlled limited-slip differential, and the benefits it adds elevate the GT3's ability far beyond what the 997 could achieve.

There's a 'but' with the 991, though, and the silver 991.2 car answers it. Unusually for Porsche, it admitted a mistake, or at least conceded to customer demands for a manual GT3, and perhaps, too, offered a concession to those customers who couldn't get their hands on the limited-run 911 R. The stick and pedal returned with the Gen2 car, which also saw its power increase via a now 4.0-litre unit to the same heady 500hp of its 991.1 RS relation. That so many still pick the PDK over the no-cost manual vindicates Porsche's addition of PDK to the GT3 mix, though it speaks volumes about Porsche's commitment to drivers that the manual made a return.

In Touring guise here – a new, no-cost package for those who specced their GT3 with a manual gearbox – it's a hugely appealing prospect. It has all the ferocity and incredible response of the bewinged GT3s, but with a visual restraint that lets it fly under the radar. Jumping between the two 991s demonstrates just how much the transmission differentiates the drive: the Touring needs all of your attention, all of your time, the PDK car allowing you some respite should you require it. They're different, but still good, both incredible, both being pure GT3 in character, performance and, crucially, feel.

It's a modern GT3, but one that still challenges and astounds whether you're wringing out every last rpm or using the engine's mid-range push and short shifting in traffic. The 991 GT3 makes either special, which underlines that at its core it remains an exemplary, and almost uniquely driver-focused car, which despite embracing technology in its current incarnation does so without running out of its analogue appeal. We'll have to wait and see what the GT department comes up with next, but if history is any nod, then it's safe to predict more brilliance from the most evocative number and letter combination Porsche makes, now celebrating its 25th birthday. **911**

TOP Return of the manual: the 991.2 showed Porsche listened to its customers who wanted the purity and engagement of stick shift over the 991.1's paddle-oriented PDK system

Total 911 verdict

996.1
- ⊕ The original. Pure, beautiful steering. A simple driver's car that's seriously undervalued presently.
- ⊖ The front-end response can be unnerving if you're not used to it.

996.2
- ⊕ More mid-range urge, neater looks and a slightly improved turn-in. Incredible to drive and unhindered by electronics.
- ⊖ Try finding one as good as this example.

997.1
- ⊕ A sharper, more purposefully styled GT3 with the ability to match its incredible performance.
- ⊖ Interior aged by the 'modern' equipment, and traction control indicates the technological creep.

997.2
- ⊕ That engine, that gearbox, that noise. Peak GT3 for the purists.
- ⊖ Purist, yes, but there's PSM in the mix.

991.1
- ⊕ Incredible ability and agility, and more rounded appeal. This is a GT3 you can drive everyday.
- ⊖ Is a GT3 a car that you should be driving every day?

991.2
- ⊕ Old and new worlds collide – this Touring is just glorious.
- ⊖ We'd love a Touring with rear seats…

Model	996.1 GT3	996.2 GT3	997.1 GT3	997.2 GT3	991.1 GT3	991.2 GT3 (Touring)
Year	1999-2000	2003-2004	2006-2007	2009-2010	2013-2015	2017-2018
Engine						
Capacity	3,600cc	3,600cc	3,600cc	3,797cc	3,800cc	3,996cc
Compression ratio	11.7:1	11.7:1	12.0:1	12.0:1	12.9:1	13.3:1
Max power	360hp @ 7,200rpm	381hp @ 7,400rpm	415hp @ 7,600rpm	435hp @ 7,600rpm	475hp @ 8,250rpm	500hp @ 8,250rpm
Max torque	370Nm @ 5,000rpm	385Nm @ 5,000rpm	405Nm @ 5,500rpm	430Nm @ 6,250rpm	440Nm @ 6,250rpm	460Nm @ 6,000rpm
Suspension						
Front	Independent; MacPherson struts with combined coil springs and dampers; anti-roll bar	Independent; MacPherson struts with combined coil springs and dampers; anti-roll bar	Independent; MacPherson struts with combined coil springs and dampers; anti-roll bar	Independent; lower wishbones; MacPherson struts; anti-roll bar; PASM	Independent; MacPherson struts with coilovers; anti-roll bar; chassis bearings partially with ball joints; PASM	Independent; MacPherson struts; anti-roll bar; selected ball joints; PASM
Rear	Independent; multi-link; anti-roll bar	Independent; multi-link; anti-roll bar	Independent; multi-link; anti-roll bar	MacPherson struts; semi-trailing arms; anti-roll bar; PASM	Independent; multi-link; MacPherson struts with coilovers; anti-roll bar; chassis bearings partially with ball joints; PASM	Independent; multi-link with helper spring; anti-roll bar; selected ball joints; PASM
Wheels & tyres						
Front	8x18-inch; 225/40/ZR18	8.5x18-inch; 235/40/R18	8.5x19-inch; 235/35/R19	8.5x19-inch; 235/35/ZR19	9x20-inch; 245/35/ZR20	9x20-inch; 245/35/ZR20
Rear	10x18-inch; 285/30/ZR18	11x18-inch; 295/30/R18	12x19-inch; 305/30/R19	12x19-inch; 305/30/ZR19	12x20-inch; 305/30/ZR20	12x20-inch; 305/30/ZR20
Dimensions						
Length	4,430mm	4,435mm	4,445mm	4,460mm	4,545mm	4,562mm
Width	1,765mm	1,770mm	1,808mm	1,808mm	1,852mm	1,852mm
Weight	1,350kg	1,380kg	1,395kg	1,395kg	1,430kg	1,413kg
Performance						
0-62mph	4.8 seconds	4.5 seconds	4.3 seconds	4.1 seconds	3.5 seconds	3.9 seconds
Top speed	188mph	190mph	192mph	194mph	196mph	199mph

/ Ultimate 911 GT3 Collection

GT3 Legacy

Total 911 explores Porsche's GT3 bloodline with a fast road drive of the first and latest homologation specials

Written by **Lee Sibley** Photography by **Ali Cusick**

GT3 Legacy: 996 v 992 | 23

It's easy to forget just how much change has taken place at Porsche in the last quarter of a century. In that time, the 911 lineup has flourished both in the number of different models available as well as numbers produced – comfortably more 911s have been made in the last 25 years than the 33 beforehand.

Porsche also diversified its products by introducing the Cayman, Cayenne, Panamera, Boxster and Macan, as well as the Carrera GT and 918 supercars; then it opened eight new experience centres worldwide, a new factory at Leipzig (with a micro factory currently being assembled in Malaysia); got taken over by VW; and started making electric cars, beginning with the Taycan. Pretty much the only constant has been the location of Porsche's HQ at Zuffenhausen, where the 911 continues to be built, albeit alongside said Taycan.

Even at Zuffenhausen though, a new era was being ushered in precisely 25 years ago: its 911 was completely redesigned from the ground up for the first time since the model's inception in 1963. A new car, the Boxster, would be built alongside it, sharing a variation of the same engine – an engine that would be cooled by water rather than air. Today, these facts are immaterial to the enthusiast. A quarter of a century ago, the move was seismic.

Two short years later, the company was also saying hello to a new age for its Motorsport cars with the fabled Carrera RS expunged in favour of a model sporting just two letters and a number: GT3. In the first instance, Porsche's new car from Weissach would homologate the 996 to go racing, in the best traditions of a Porsche sports car. But it would do much more than that: the first 'hot' 996 would start a deep love affair between devout driving enthusiasts and pretty much anything to roll out of what is now known as the 'GT Department'. Some 25 years and the production of 20,000 models over seven generations later (not including RS), the 'GT3' appellation might well be the most evocative in the Porsche lexicon.

The latest and greatest of those is of course the 992 GT3. We have previously tested it on its customary proving ground, the race track. But the magic of the GT3 as a homologation special is that while its skills are honed for the circuit, it's also applicable to street driving.

Around town, there's no question it's an easier proposition than the 991.2 it replaced: the clutch and shifter are much lighter and easier to engage – even from cold – and mechanical chatter from the clutch release bearing is less perceptible in the cabin, which also filters out a degree of rolling tyre noise too. It's not as agricultural in urban areas as the 991.2, which makes daily driving a semi-realistic proposition, though a stiff ride should still deter all but the most committed of enthusiasts.

That's not to say the GT3 experience has been diluted in the latest offering. Even at low speeds, its front axle feels sharper, and whereas in the 991 you had to be hitting the upper echelons of its maximum 9,000rpm for the motor to really sing, here the flat six symphony begins in earnest from as little as 5k. Clearly, the bandwidth of the GT3's performance has been increased, but great roads are where it pleases most, which is why we find ourselves in the middle of nowhere on the North York Moors, with an old friend in tow. A fun, smooth(ish) network of roads running mostly in parallel over these picturesque moors, they are simply a paradise for those who

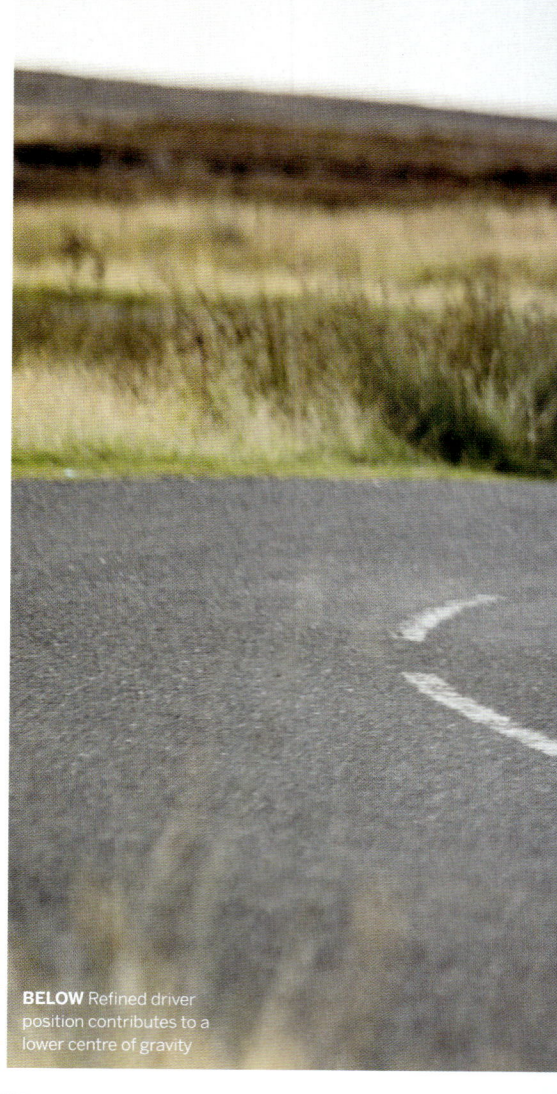

BELOW Refined driver position contributes to a lower centre of gravity

BELOW Swan neck strut design is a first on a road 911 and generates 50% more downforce over 991.2 (combined with diffuser)

GT3 Legacy: 996 v 992 | **25**

Model **992.1 GT3**
Year 2020

Engine
Capacity 3,996cc
Compression ratio 13.3:1
Maximum power 510hp @ 8,400rpm
Maximum torque 470Nm @ 6,100rpm
Transmission Six-speed manual

Suspension
Front Independent; double wishbone with anti-roll bar; all chassis mounts ball joints; integrated helper spring; PASM
Rear Independent; multi-link; anti-roll bar; partial chassis bearings with ball joints; integrated helper spring; PASM

Wheels & tyres
Front 9.5x20-inch; 255/35/ZR20
Rear 12x21-inch; 315/30/ZR21

Dimensions
Length 4,573mm
Width 1,852mm
Weight 1,418kg

Performance
0-62mph 3.9 secs (manual)
Top speed 199mph

"On a public road, the barometer of accomplishment is smiles, not sector times, which puts the 996 eye to eye with its distant, exuberant relative in the 992"

GT3 Legacy: 996 v 992 | 27

28 | Ultimate 911 GT3 Collection

ABOVE 996.1 GT3 actually weighed 30kg more than an equivalent C2, but boasted a dry-sumped 'Mezger' engine, uprated chassis and bigger brakes

Model **996.1 GT3**
Year 1999

Engine
Capacity 3,600cc
Compression ratio 11.7:1
Maximum power 360hp @ 7,200rpm
Maximum torque 370Nm @ 5,000rpm
Transmission Six-speed manual

Suspension
Front Independent; MacPherson strut; anti-roll bar
Rear Independent; multi-link; anti-roll bar

Wheels & tyres
Front 8x18-inch; 225/40/R18
Rear 10x18-inch; 285/30/R18

Dimensions
Length 4,430mm
Width 1,765mm
Weight 1,350kg

Performance
0-62mph 4.8 sec
Top speed 188mph

love driving – and my current predicament is equally as heavenly.

The view ahead consists of a winding road twisting excitedly over quiet moorland; the view behind is dominated by the swan neck uprights of the 992's wing but behind that, a flash of red reveals Neil Plumpton's 996.1 GT3. Separated by 22 years, these two homologation specials neatly bookend Porsche's GT3 story – and as we've just discovered, much has changed at the company in the interim. So is it reflected in the cars that share a name?

Our drive of the 992.1 Touring in issue 208 revealed the 992 GT3's front axle to be a quantum leap over the 991, so quick and sharp is the car at its nose. However, there are differences – albeit subtle – between the Touring and this track-oriented, winged equivalent. Its ride is slightly stiffer, for one, and the steering wheel is noticeably busier over bumpy blacktop too, requiring something of a wrestle to keep the car from tramlining all over the road's divots.

Another big difference to that drive in the Touring concerns the GT3's transmission. Our Touring on test was PDK, whereas this winged GT3 has Porsche's six-speed Motorsport gearbox – and what a revelation it is. Sporting the lightest pedal of any Porsche GT car I've driven, the shifter too is fairly weightless (but not superfluously so), offering a short, precise throw. It's simply a joy to use and is the best ally yet to the brilliant 510hp, 4.0-litre flat six stuffed out back.

My only gripe is pedal positioning. It's less than ideal in the 992, the accelerator's organ pedal placed too far over and with too much of an offset to the brake pedal to facilitate seamless heel and toe. There is a cheat though, as selecting Sport mode sees the car automatically rev match to keep the GT3 smooth between downshifts, while on the way back up, the GT3 reveals another neat feature of its arsenal: flat shifting. In the 992 there's no need to lift off the throttle when changing up a gear, thereby reducing the amount of time the 911's rear wheels are starved of power. Instead, with my right foot planted to the floor, a kick at the clutch with my left foot and a swift slide of the shifter sees the GT3 accept a deft gear change in rapid time. That, and the fact it can tap into an additional 100Nm of torque, means the thunderous 992 can muster a sizeable margin of space to its distant forebear in a straight line, though in corners the latest GT3 has the ability to dust off the 996 completely.

The 992's mechanical grip is simply astonishing. Unrivalled among other road-based 911s, the sheer speed I'm able to carry through bends is frankly ludicrous, and yet the GT3 never feels like it's even remotely beginning to tickle the limits of adhesion. No matter how hard I push, or how bumpy the road becomes, this GT3 refuses to break contact with the floor, which is evidently a key factor behind its ability to cover ground so quickly and with such assurance.

After an hour or so of playing cat and mouse across the moors, we pull over for a debrief, and Neil – a serial owner of more than 30 Porsche in the last 16 years – is equally stunned by the 992's grip. "That thing's mad," he says, pointing at the Shark blue car. "There were a couple of times you went into a ➲

ABOVE The 996 GT3's cabin is indicative of a time when simplicity was king, with no buttons or toggles to activate driver modes or chassis aids

corner with such speed, I thought surely you'd be in trouble, but it just gripped and gripped. I can't believe it." It's an an apt appraisal indicative of the 992 GT3's sizeable capabilities, rather than those of the wally in charge of steering it.

Before Neil and I swap cars, we take in their profiles on the hillside. Even when merely parked next to one another, it's hard to believe these two 911s share much more in common than the model denomination stamped to their backsides. The sculpted 996 is small and dainty compared to the jagged 992 with its more aggressive styling.

Aesthetically, the 996 is only a minor rework over its Carrera sibling: still narrow-bodied and with exhausts remaining faithful to the 911's rear corners, there are none of the extra vents that would appear ahead of the bonnet and on the rear PU from the 997 onwards either. Even the bi-planed 'taco' wing sitting atop the decklid feels reassuringly minuscule compared to the 992's swan-necked approach to modern-day aerodynamics.

The 996 does sit 30mm lower than its C2 sister though, plus there's a raft of changes you don't immediately see like adjustable dampers, stiffer springs, bigger brakes, plus a 993 GT2-spec gearbox and dry-sumped 'Mezger' flat six in the back.

Stepping into the 996, it really does hit home just how much the GT3 has evolved in the years under Mr Preuninger's tutelage. This first GT3 may have heralded the modern era of Porsche GT car, but to experience it really is like going back in time. The simplicity of its cabin is indicative of the 996's general approach to fast motoring: there's no Sport button, let alone a Mode wheel as part of any Sport Chrono Package. As for the Sport setting of Porsche Stability Management, with its greater slip angle? The only tools for mitigating traction here are your feet and hands.

I'm sat far higher up in the 996 too (though the Clubsport leather buckets offer an excellent hold), the steering wheel placement feels awkward, the only adjustment being a forwards/backwards movement, and the manual shifter feels like it's placed somewhere down near my knees. It shows Porsche really has honed the driver's seating position to perfection in the years since.

We venture back out on the same roads we've just explored, where it takes just seconds for Porsche's first GT3 to endear itself to me. Vision, first of all, is first class: I can see all the 996's extremities from the driver's seat, which makes placing this stunning, 35,000 mile example on the road a far easier task than in the bulky 992. Its pedals too are ideally placed and offer the perfect tools for quick heel and toe, even if the clutch has such weight to it that it feels akin to a gymnasium leg press.

The wheel is jostling about in the palms of my hands, and the chassis is already showing itself to be playful as we move through a succession of medium-speed bends. Building pace, the M96/79 boxer shows plenty of character and eagerness to rev to its 7,600rpm redline as I try and chase down the 992. Before long we reach a fast, darting S-bend, where earlier the 992's lightning-quick directional changes and ability to maintain grip had left both Mr Plumpton and I dumbfounded.

Needless to say, the 996 will not carry anywhere near as much velocity into this first right-hander, and so while the 992 monsters on, I'm hitting the 996's brakes to scrub speed for an acceptable corner entry. Turning in, it takes a while for the car's nose to follow my inputs at the wheel by comparison to the razor-sharp 992, but the process in this 996 is nevertheless beautiful: you feel so much more of *everything*, from rear to front load, to lateral weight transfer, to the (at times) teetering levels of grip being eked from those much skinnier Pirelli tyres.

The 996 darts right, I then lift slightly, push the wheel left and feed the throttle in, before accelerating hard out of the S-bend… wow! This GT3 feels alive in my grasp, and at considerably less speed than the 992 needs to muster to deliver the same feeling. The 996 just feels so playful: whereas in the 992 you never quite feel in sole charge of the car, the 996.1 lets you take it by the scruff of the neck, giving you complete autonomy to find yours or the car's limit, whichever comes first. The experience rewards on a whole new level, and really highlights why many view this pre-computer era as the high watermark of sports cars, if not in terms of their technical prowess but the sheer engagement they offer. So how to compare the first and latest Porsche GT3?

Really, there's no comparison to be had between the 996.1 and 992. They hail from completely different eras, and it shows. There's an innocence to the 996 in its approach to performance, yet the 992 is simply a technical masterpiece, the result of more than two decades of obsessive attention to every gram of weight, every drop of petrol, and every molecule of air that goes near it. The 992 GT3 is a race car with a sat nav, whereas the 996 is simply a classic sports car.

However, on a public road, the barometer of accomplishment is smiles, not sector times, which puts the 996 eye to eye with its distant, exuberant relative in the 992. On this metric alone, choosing a champion is difficult.

It's hard to look past the sheer theatre of the 992, but as we get our final shot at sunset to wrap up an entertaining day on the moors, it doesn't stop me from asking Neil if I can drive his 996 back to base for a late evening supper. There's just something about its purity through simplicity that I'm drawn to, and I'm keen to savour it one last time.

And that's how our day ends: two GT3s chasing down the very last remnants of sunlight en route back to civilisation, the 996 and I following the warm glow of the 992's full-width rear light bar as it dances in the darkness across the deserted moorland roads.

Where to from here? Only Mr Andreas Preuninger really knows for the short term, but one thing's for certain: in another 25 years' time, Porsche's GT3 will be another very different beast indeed. 911

GT3 Legacy: 996 v 992 | 31

32 | Ultimate 911 GT3 Collection

997 Rennsports

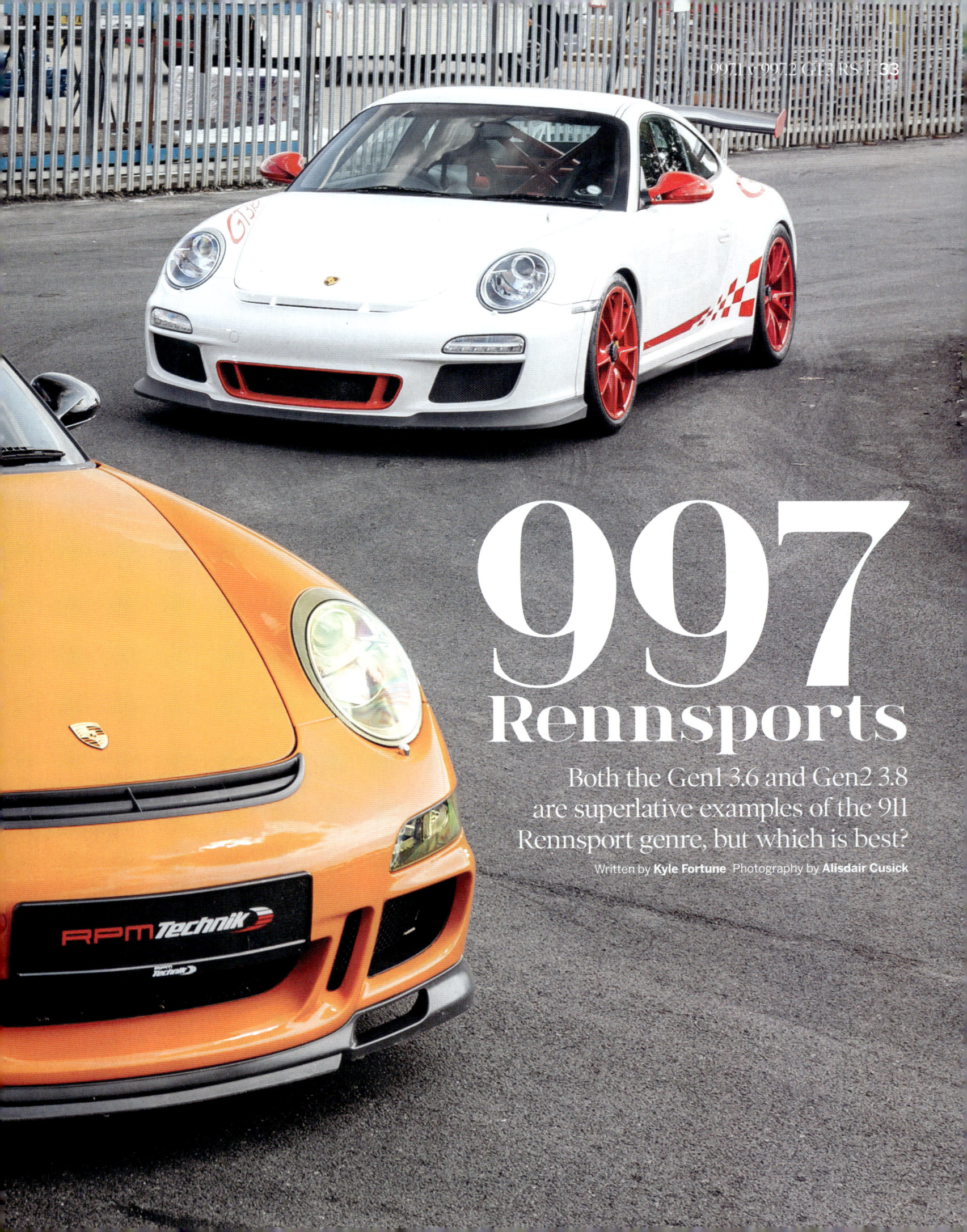

Both the Gen1 3.6 and Gen2 3.8 are superlative examples of the 911 Rennsport genre, but which is best?

Written by **Kyle Fortune** Photography by **Alisdair Cusick**

Ultimate 911 GT3 Collection

It was the 996 that started the GT3 RS dynasty, fusing a new GT3 platform with those hallowed RS letters which have long signified a Porsche to be significantly special. That first liquid-cooled RS was something of a toe-in-the-water for Porsche. The management was sceptical that the Andreas Preuninger-led, after-hours built, overt (for then) GT3 RS would find customers. It did, and convincingly so, with Porsche unable to make enough of them for the demand that it created.

So, these two cars, and all of us who enjoy driving, have a lot to thank that 996 for, because without it the 997 might never have been offered as an RS. That designation was instrumental in Porsche realising the commercial value of these focused, enthusiast cars, not just as another model line to sell, but as a tool to bring credibility to the rest of the 911 range as a direct link to the company's racing activities.

What's almost certain is that while Porsche was struggling to sate the appetite of customers wanting the 996 GT3 RS, the 997 GT3 RS was already under development. The GT3 it was based on was shown at the Geneva Motor Show in March 2006. No management concerns this time, with Porsche taking very little time to follow its new GT3 with an RS version. The 997.1 GT3 RS debuted at the 2006 Paris Motor Show, or correctly 'Mondial de l'Automobile', along with the 997 Targa 4S, which couldn't be much further removed from its track relation.

There would be one similarity, though: the adoption of the wider body from the four-wheel-drive cars that saw the rear arches swell by 44mm over the slimmer-hipped GT3, for the wider track and, as it was described back then, "transverse acceleration potential". That's cornering speed in case you're wondering. Obviously, it remained rear-wheel drive, with the 3.6-litre Mezger engine's output hitting 415hp at 7,600rpm (800rpm short of its 8,400rpm redline).

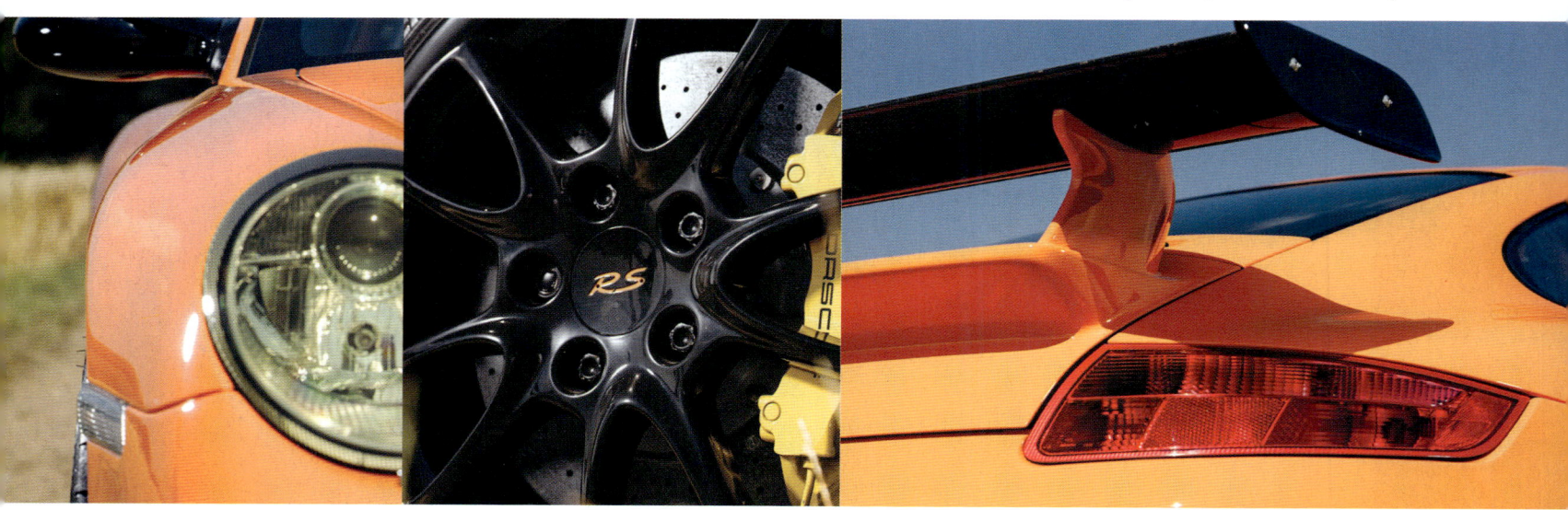

ABOVE The 997.1 GT3 RS made its debut in the autumn of 2006 at the Paris Motor Show. It was originally available in four colours: Arctic silver, Black, Viper green and finally Pure orange, as shown here. The 19-inch alloy wheels are painted black, as is part of the rear wing and wing mirrors, helping to create focal points that draw the viewer's eye across and over the car

The high-revving motorsport engine was lifted from its GT3 relation, itself essentially the flat six from the 996 GT3 RS before it. I remember receiving the press release when the 997.1 GT3 RS was announced. It was added to the release about the GT3 with a scant, almost apologetic eight paragraphs and without much in the way of detail. Greenlit as a model or not, it was evident that there was a bit of a hangover within the company as to how to communicate its most focused 911 development for road and track.

No such issues in the GT department where, rather than the somewhat racer homologation look of its 996 GT3 RS predecessor's white-only body, the 997.1 GT3 RS came with its warpaint already applied. Without going down a wormhole of rarities, or special orders and requests from favoured customers, or the "my mate's got one in…" claims, the 997 GT3 RS was officially offered in four different hues: Arctic silver metallic or black, or as an option in orange or green. To that the "vehicle insignia and the wheels themselves are styled in orange or black to contrast with the body paintwork", to use Porsche's then press release parlance. Those optional colours would prove popular. When was the last time you saw an Arctic silver or black 997 GT3 RS? They do look wonderful when so specified, mind.

In case you missed it in the photos, the car we're in today was optioned by its first owner in that orange, which means black wheels and decals. It looks as good now as it did when it was new. The GT3 RS suits strong colours, and there are few more vivid than here. Not so much that it doesn't enable you to drink in the details. The front is far more obviously sculpted to channel air into, around, under and over it, with the signature vents atop that front bumper now looking like they were always meant to be there, rather than the somewhat jigsaw-cut-out-during-a-hot-race look of those on the simpler-lined 996 GT3 RS before it. ⤴

ABOVE The wheel arches of the 997.2 GT3 RS were widened to take into account the slightly longer front axle. The wheels differ from the Gen1's, and are now of the central locking variety. The rear wing is wider, and sits on taller uprights. The model for sale at RPM Technik was optioned in Carrara white by its original owner, with red accents across the bodywork

36 | 911 GT3 Collection

ABOVE AND BELOW
Subtle changes to the front of the 997.2 GT3 RS better manipulates air into, around, under and over the car, improving its aerodynamic performance and downforce

Indeed, it's difficult to comprehend that just three years or so separate this car and its ancestor. Such is the pace of development – particularly when that's accelerated with the desire to race – that the 997 GT3 RS looks far more aggressive than that car and any of its GT or RS predecessors. Time and its ever-more pugnacious successors has dulled that feeling slightly, but it's still a car that unashamedly exhibits its intent. The body might have been wider, but thanks to the usual RS diet the 997.1 GT3 RS weighed in at around 20kg less than the standard GT3. It didn't go to quite the lengths of its predecessor to achieve that, with this car featuring an aluminium bonnet rather than the carbon-fibre one of the 996 RS. Similarly, this car's badge was exactly that: a badge rather than a sticker.

There is carbon fibre in its make-up, though, with the adjustable rear wing being made of the black weave (as are the bucket seats inside), while the rear windows were plastic, with a good deal of the sound deadening also binned in the pursuit of weight loss. Other key changes include adjustable suspension elements, with that suspension now coming with electronically adjustable dampers, while the GT3 RS also gained traction control with either 'on' or 'off' settings. If you want a bit more freedom with it on, push the Sport button. Doing so not just ups the threshold of the traction control, but opens the exhausts for significantly reduced back pressure that increases the torque slightly, too. Two buttons… that's all the adjustability you've got, unless you want to get your hands dirty and start messing around with shims on those split arms.

You're unlikely to want to, because out of the box the GT3 RS feels sensational. It's something that's not dulled one bit since the days where we could call up the UK Porsche press fleet and borrow one. The 3.6-litre, with its quoted 415hp, has a keen-edged ferocity that's largely a result of its significantly lightened single mass flywheel. The gearshift has some heft, particularly before it's got some heat into it, but the action through the gate is a real delight.

Sitting in the cabin, you're embraced by the clutch of the brilliant seats, with the cage criss-crossing the interior – that cage featuring more bracing than that in a GT3 Clubsport for greater stiffness. The ⮕

"It's a supremely capable, ridiculously engaging point-to-point car that takes everything you give it and keeps asking for more"

38 | Ultimate 911 GT3 Collection

> "On the road it feels every bit as fantastic as it always did… this feels and looks 'RS', as it should"

steering wheel is covered in Alcantara – a nod to Preuninger's love of the Lancia Delta Integrale. It feels and looks RS, as it should. If you're used to the genre today there are some things missing. The door cards feature proper handles rather than pulls, and similarly, outside there are five lugs holding on each of the wheels, behind which sit PCCB brakes, which were still an optional item on the RS when new.

On the road it feels every bit as fantastic as it always did. The 997.1 GT3 RS is so faithful to inputs, fast and surprisingly fluid given the focus of its suspension. The engine's ample low-rev urge underpins a maniacal quest for revs that's difficult to ignore whenever the opportunity arises. The steering is a step on from its predecessor, with the 997.1 GT3 RS's nose turning in with greater authority and holding on for longer, and here the rear axle aids with mighty grip and similarly impressive traction.

It's a car that Preuninger is particularly fond of, explaining that when he would visit his wife and new-born son in hospital, he did so in the then-new 997 GT3 RS. It was the car he actually brought his lad home in. His son apparently doesn't know that, but when asked what's his favourite GT car from his dad's greatest hits, he's drawn to that Gen1 997.

All this is against the general consensus: that the best RS is the next RS. That's irrefutable if you measure things like lap times, with the 997.2 GT3 RS bettering the 7:42 that Walter Röhrl achieved around the Nürburgring in its predecessor by around nine seconds. That doesn't really come as a surprise when you consider that the 997.2 GT3 RS's goal was to create a wider gap between it and its 'regular' GT3 relation. It's got 15hp more power than the GT3, thanks to a combination of a development of the now 3.8-litre Mezger that hung out the back of the GT3.

Like the Gen1, it revved quicker because of the fitment of a single mass flywheel, which is 8kg lighter than the regular GT3, and 1.4kg lighter than that of the Gen1 997 GT3 RS. The maximum engine speed rose to 8,500rpm, with the maximum 450hp produced at 7,900rpm, and peak torque of 430Nm coming at 6,750rpm.

The Sport button again adds some more torque into the mix, with an additional 35Nm produced at lower revs. This gain doesn't add to the overall ➔

ABOVE AND BELOW
Both cars are fitted with a roll cage and bucket seats, with carbon fibre used to reduce weight inside the driver's cabin. The use of Alcantara on the steering wheels underlines the sports heritage of Porche's esteemed RS line-up

40 | Ultimate 911 GT3 Collection

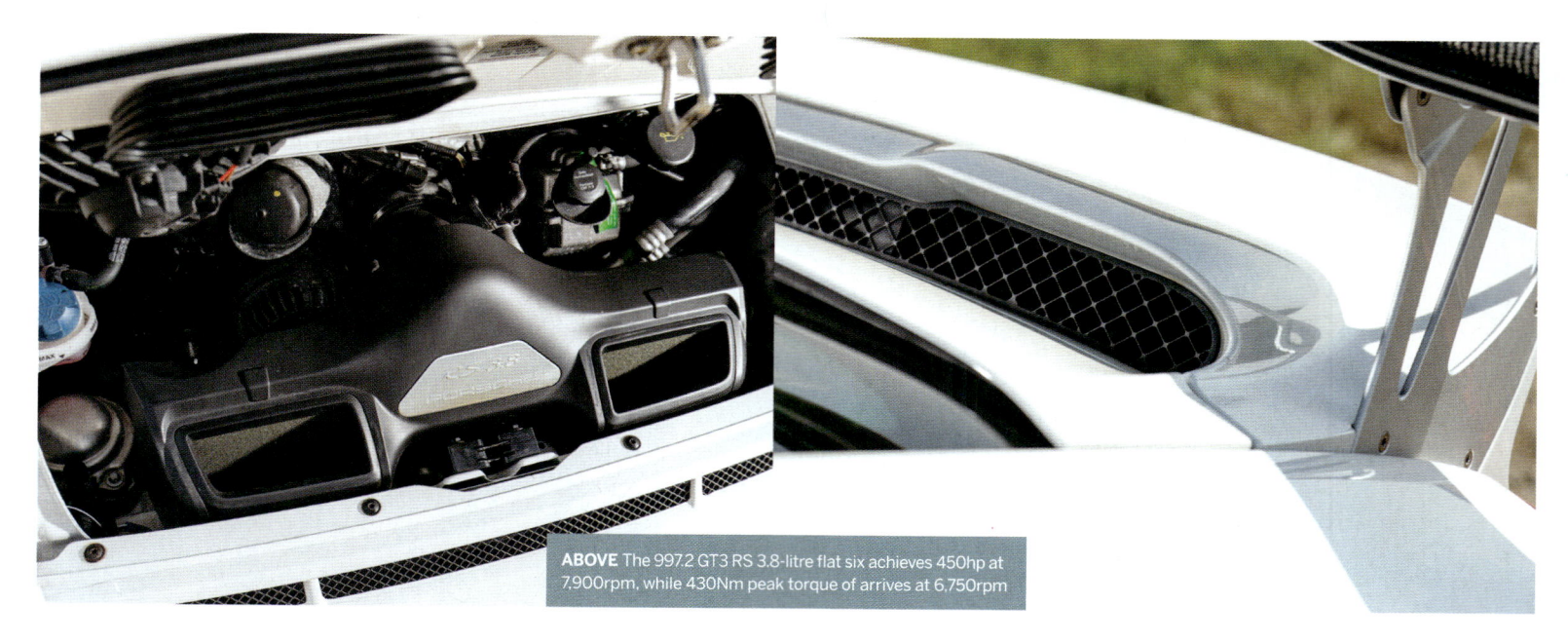

ABOVE The 997.2 GT3 RS 3.8-litre flat six achieves 450hp at 7,900rpm, while 430Nm peak torque of arrives at 6,750rpm

peak output, but increases flexibility. It's quicker, but today we're not crunching the 0-62mph or top speed numbers. They really don't matter. What does is how the cars feel, and both are fantastic.

To the Traction Control of the Gen1, the Gen2 benefitted from PSM (Porsche Stability Management), which offered the possibility of being switched off, or the intervention levels heightened with that Sport button pressed. Significantly, the Gen2 car gained a bit of track width on the front axle, necessitating some add-on wheel arches. The whole exterior look of the Gen2 RS becomes even more in your face than before. The decals, once relatively respectfully kept at wheel centre level, kick up over the rear arch on the Gen2, with the front wing gaining an evocative GT3 RS script, too.

Inside the car, you feel not just the march of that technology and those driver aids, but the involvement of the marketing people in the specification, with things like those pull-straps instead of the Gen1's proper handles. It gains a more blatantly racer schtick: these straps accepted as the norm now but in reality they add little and – if we're being brutally honest – don't work as well as the handles in the Gen1 do.

Along with the showier decals, that bigger rear wing (placed higher on cool-looking metal struts) as well as details like the centre-lock wheels, the 997.2 GT3 RS is arguably the best looking of the RSs. It balances its purposefulness while still retaining a

ABOVE The 997.1 GT3 RS has a less powerful 3.6-litre engine than its replacement, but can still achieve 415hp

degree of purity. Just park one alongside the very busy-looking 992 GT3 RS and you'll understand what I'm saying. It has for a long time, along with the 996 GT3 RS, been a favourite of mine. I attended a launch event for it back in 2009, where a drive of a grey and gold example on the roads climbing up from the back of Nice, France, on roads that the Monte Carlo rally is held, remains a career highlight. The opportunity, then, to drive a Gen2 RS is never one that I'll pass up, or ever be disappointed with.

And so it proves: the 997.2 GT3 RS is sensational. It's a supremely capable, ridiculously engaging point-to-point car that takes everything you give it and keeps asking you for more. It's an incredible car, feeling more polished and more complete than the orange car here. It rides with remarkable composure, that front axle even more predictable and precise. The Gen2's engine is an absolute screamer, too, and the gearshift is similarly brilliant – the engagement that it delivers so rich and pure.

Mightily impressive as it is, it doesn't necessarily make the 997.2 GT3 RS better. Indeed, there's a bit of me that yearns for the slightly rawer, edgier experience that the earlier Gen1 Rennsport delivers, the way it demands a bit more of your attention even, and especially when travelling at lesser speeds. That's pretty revelatory, because had you asked me prior to driving these cars back-to-back, I would have unequivocally singled out the 997.2 as the GT3 RS you need to buy, at all costs.

42 | Ultimate 911 GT3 Collection

It is, but regardless of all the reasons that make it so brilliant, and quantifiably better than its forerunner, there are elements of the earlier car I think I prefer. Preuninger's son might just be on to something. That's not 'better' necessarily, but for me, and today, the Gen1's the more compelling car. It's a close-run thing and on a different road, changeable conditions, or throw a track into the mix, and it'd possibly – and likely – change. Even so, the Gen1 is the one I've enjoyed the most today, and that's before even considering the sizeable price differential between them (enough to have the Gen1 and another 911). Both these cars are currently on sale at RPM Technik. Go check them out. They're both superb examples of incredible cars, but perhaps go in with one in mind and with intent to buy, because if you don't, drive both and you may just end up changing your mind, repeatedly… **911**

> "The 997.1 GT3 RS is so faithful to inputs, fast, and surprisingly fluid given the focus of its suspension"

Total 911 verdict

There's no such thing as the wrong choice when deciding on buying a 997 GT3 RS, because they're epic drivers' cars. The differences are arguably bigger than between any other RS generational revisions, but that doesn't preclude the desirability of the earlier car and, depending on your outlook, it might actually make it more appealing. Get either, now, before they become even more unattainable.

997.1 GT3 RS

LIKES
- Still a very pure car, looks sensational, seats feel lower and tighter than the Gen2, plus it's raw and engaging

DISLIKES
- Everyone will say you should have bought a Gen2

997.2 GT3 RS

LIKES
- Greater polish and poise, still an incredibly exciting, involving and ridiculously rapid car

DISLIKES
- Mad as it sounds, the greater track focus makes it slightly less appealing as a road car

White Lightening

The 996 GT3 RS is brilliant out of the box, but Porsche always intended for customers to make some changes...

Written by **Kyle Fortune** Photography by **Alisdair Cusick**

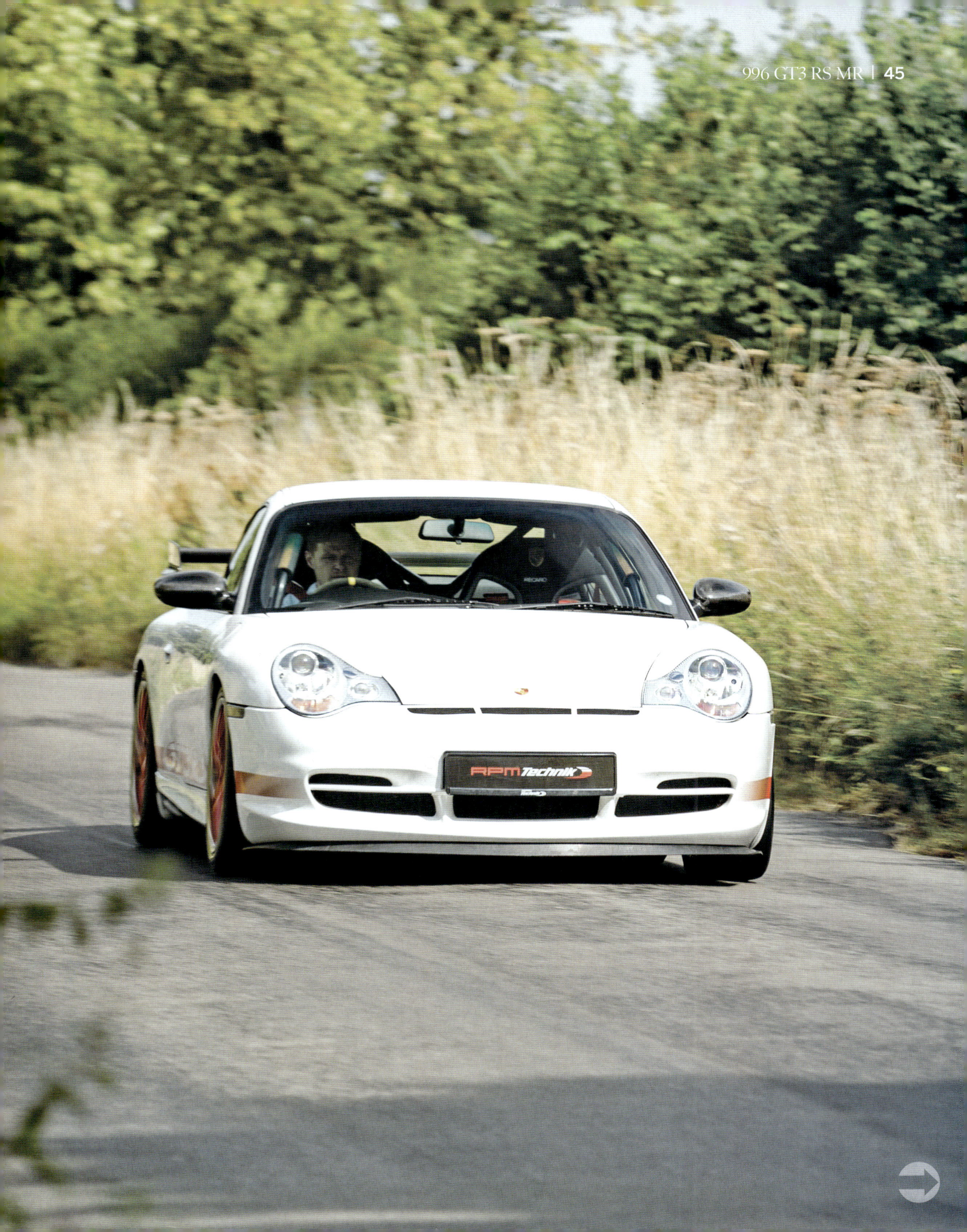
RPM Technik

Full disclosure here. The 996 GT3 RS was the first RS I ever drove and it left something of an impression – so much so that I'll always look for an excuse to drive one. That first car was owned by a friend, Neil Primrose, who had the first UK car in the country back in 2003. It arrived before the UK press car and he was kind enough to let me borrow it for a day. This experience, coupled with subsequent drives of that delayed UK press car, Porsche's own museum car and other privately owned examples over the years, have only reinforced my opinion that the 996 GT3 RS is among the greatest of all RSs. If you truly covet purity and want to enjoy your car on both road and track, the 996 RS is a hugely compelling machine.

Inevitably then, while checking out the RPM Technik website for the specifications of the two 997 GT3 RS that appear elsewhere in this issue, this 996 GT3 RS caught our eye. Not just because at £149,995, like all 996 GT3 RSs, it's criminally undervalued in the current marketplace, but because it's a car that's been modified. Originality obsessives can stop reading, but my interest was suitably piqued. Similarly, if you subscribe to the RS philosophy then this car should also be of interest to you.

All RSs add the possibility of adjustability to the 911s they're based on (easier today with a 992.1 GT3 RS thanks to steering wheel dials). The expectation – desire, even – is that you'll rejig things to suit you and your intended use of it. If that means modifying it, then so be it. These are legitimate near-race cars after all, and meant to be enjoyed as such. If doing so, then where better than to go than Manthey Racing? It's a company that's so intrinsically linked with Porsche and the GT department that not long ago Porsche formalised the relationship by taking a 51 per cent controlling stake in the business.

As a true homologation car the 996 GT3 RS was spun off its already impressive 996.2 GT3 relation, as a product of necessity to improve Porsche's customer racing car. The project was one of secrecy, too: GT boss Andreas Preuninger admits it was carried out without management knowledge. These kind of skunkworks, after-hours projects built by enthusiasts for enthusiasts always create something very special. And so it proved with the GT3 RS.

The story gains even more romance due to the fact that Preuninger based the car's look on his boyhood poster car, the 2.7 RS. The 996 GT3 RS is the manifestation of those childhood dreams, and specifically, young Preuninger's love of a blue decal on a white '73 car. 'His' RS would be offered in either that combo, or like here, red over the Carrara white paint. Of the two, red is more common, with around two thirds said to be so specified, and while there's no official number for blue over white we've heard "181" uttered among people who should know…

Even with that reverential nod to Porsche royalty, Preuninger and his team had to fight to secure board approval. Many of those within country marketing teams were sceptical that such an overt-looking car – this being 2003 – would find customers. Thus, the projections for sales were pessimistic and some markets, notably the USA and Canada, weren't even offered it. The original press release stated it would only build around 200, though latterly Porsche said it anticipated a maximum build number of 400. Of that 400, a sizeable figure of 99 cars were pencilled in for UK buyers. However, that would increase to 114 C16 UK cars at the end of production.

Indeed, the UK would have taken even more. It was vindication that the GT department's RS experiment was justified, with demand far exceeding supply. The company could have sold two or even three times the number it would eventually produce, with that final output reaching 682. Porsche would have built more, but the original production tooling that was created for that anticipated 400 number was worn out. All this makes the first liquid-cooled RS the rarest of the series production naturally aspirated RS models, with only specials like the later 997 GT3 RS 4.0 being produced in lower numbers.

The changes made to the 996 GT3 to turn it into an RS followed the well-proven route taken by its air-cooled predecessors. A reduction in sound deadening dropped some mass and increased the soundtrack inside the pared-back, two-seater only cabin. The glass from the B-pillar back was replaced by acrylic, while the bonnet was made in carbon-fibre composite upon which, famously, the enamel and metal badge was replaced by a sticker. Those saved a few grams, along with the use of more carbon-composite ➲

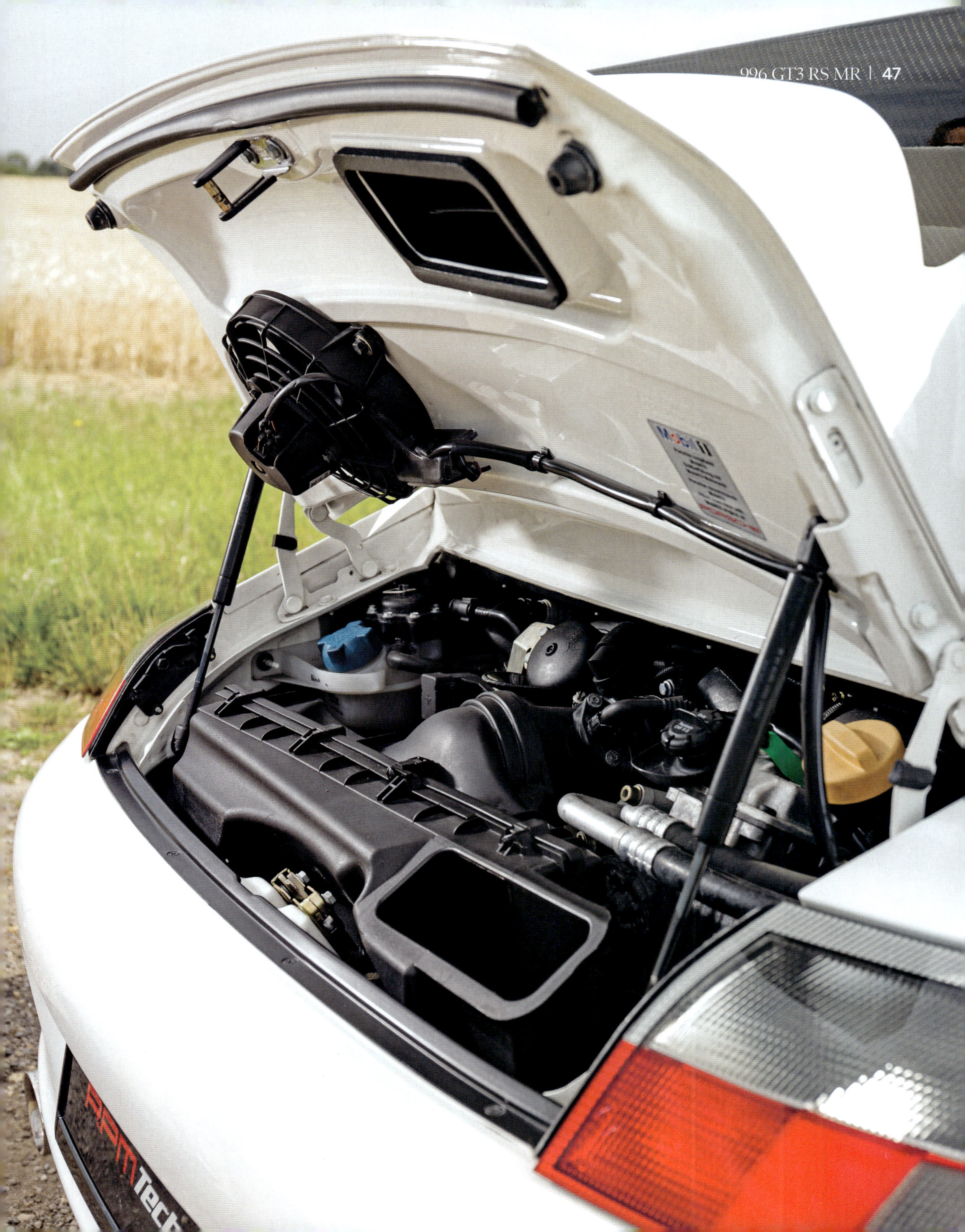
RPM Tech

48 | Ultimate 911 GT3 Collection

RIGHT The roll cage inside this Manthey Racing example hasn't been painted white, and thus makes for more harmonious surroundings

ABOVE Unapologetically raw, the 996 is the only narrow-bodied 911 GT3 RS, and the only one without any form of traction control

panels – the rear wings, door mirrors (left unpainted to show the back weave) – as well as a lighter, plastic rear-engine cover. That engine lid featured a scoop to deliver a ram-air effect to boost engine power, by around 15hp, at 187mph. The weight savings meant a kerbweight of 1,360kg, which is around 20kg less than a 996.2 GT3 Clubsport. Other visual changes that also aided performance included venting fore of that lightweight bonnet and a carbon-fibre wing that added around 35kg of downforce at 125mph.

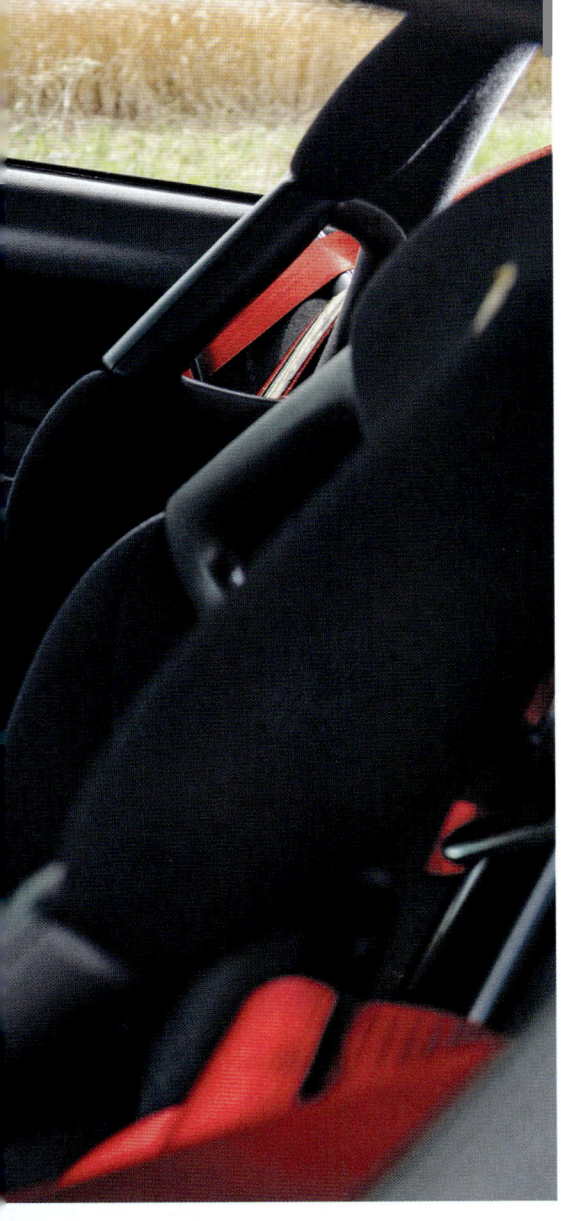

Perhaps most significantly for homologation were the chassis revisions, which brought strengthened wheel carriers that benefited camber control – these were hollow cast for weight saving. Similarly, the suspension top mounts were adjustable and strengthened, while the control arms front and rear allow adjustment, all these bits essentially being RSR parts. Attached to those hubs were new alloy wheels. The part-painted, part-polished finish was heinously expensive to productionise, but worth it, with the owners having the choice of optioning PCCBs (Porsche Carbon Ceramic Brakes) behind them – as most did – or sticking with steel items.

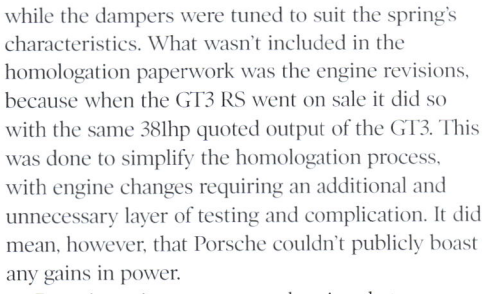

Riding around 3mm lower than the GT3, the springs and dampers were unique to the RS. The springs were progressive rate rather than linear, while the dampers were tuned to suit the spring's characteristics. What wasn't included in the homologation paperwork was the engine revisions, because when the GT3 RS went on sale it did so with the same 381hp quoted output of the GT3. This was done to simplify the homologation process, with engine changes requiring an additional and unnecessary layer of testing and complication. It did mean, however, that Porsche couldn't publicly boast any gains in power.

Preuninger is now on record saying that none of the 3.6 RS Mezger engines left the building with less than 400hp. That fact was borne out by many customers having taken them to dynos and always witnessing numbers beginning with a four. Figures of 410-415hp are often quoted. These are dyno ratings, and don't take into account the ram-air effect when running at close to maximum speed. Changes to the engine included reshaped intake and exhaust ports, which were instrumental in the greater output. The fitting of a single mass flywheel resulted in a faster, freer-revving engine. It produces its maximum power, whatever that may be, at 7,400rpm before reaching its redline at 8,200rpm.

Use all that power and it'll reach 62mph in 4.4s on its way to its 190mph top speed. Yet top speed and acceleration numbers were never really the key goals for the RS. It was more about how they were achieved and exploited. In the right hands, Porsche ambassador and occasional test driver Walter Röhrl managed a Nürburgring lap time of 7:43, which back in 2003 was very brisk. Remember, too, that the 996 GT3 RS does without any stability or traction control. It's the only GT3 RS to do without such driver aids.

It's been around 14 months since I last drove one and getting into the car at RPM Technik is a quick reminder of how basic the 996 cabin is. It's compact, too, and feels like a small car. The weather is hot, so it's heartening to see that this car was fitted with air conditioning. There's the tangle of the cage to negotiate and the racing seat belts, too, although we're on the road today and are using the inertia reel belts. That cage isn't, like many are, optionally painted in white, and it's all the better for it, because it does hide it a touch. What's obvious is the replacement of the standard air-bagged 996 RS steering wheel for a MOMO one. It has quite a deep dish and lengthy boss, which means it can be up close and personal.

You'll have to be keen-eyed to spot the Manthey changes. There's a nod on the tailpipes, which instead of being partially oval to fill the lower rear bumper cut-outs are round, with Manthey script engraved into the top of them. That's it for the visual cues, unless you've spotted the shims that allow some additional rear-wing adjustment. Those pipes make it

"It's a car that's unquestionably rawer and more intense than its unmodified alternative"

pretty clear that this 996 GT3 RS isn't quite standard when it starts up – there's an appreciable increase in bark when the engine fires. It sounds more racer in its note than the standard car, which is a good thing.

If it's clear outside it's even more so in that pared-back interior, though while it's more obvious it's not to the point of being obnoxious. It's not enough to drown out the chunter from the clutch release bearing at idle, either. That sound is always welcome because it means there's a Mezger behind you.

Sounds aside, what's clear is that there's been some revisions to the suspension, with the car riding on fully adjustable coilover suspension and a geometry setup developed by Manthey. The front axle on a standard 996 GT3 RS is more precise than its regular GT3 relation, but even so there's still some evidence of a bit of float at speed. That's eradicated with the Manthey setup, with the nose feeling absolutely planted. Grip is mighty, as is traction, both improving thanks to 21 years of tyre development, with the Michelin Pilot Sport Cup 2s this car's wearing unquestionably aiding here. The ride gets busy on the worst roads, which makes the proximity of that steering wheel helpful, but it's also transformative, and brings with it an even greater level of detail of what's going on under the front wheels.

If there's a bit of a trade-off it's a slight loss of self-centring when returning lock. This is something that you quickly adapt to, though. And the additional ➲

50 | Ultimate 911 GT3 Collection

input is well worth it, because you'll benefit from greater feel, precision and the ability to cover ground. The gearshift feels much the same as standard, taking a bit of time to warm through, but once there's heat in it the shift's wonderfully precise and quick. The small gearknob fits in your hand far more comfortably than those in the later 997 GT3 RS – even more so if you're shifting wearing racing gloves.

Improved as the steering is, there's a sizeable increaset in the performance from the engine. It's fitted with Manthey's K410 conversion – back in 2006 – which is said to add around 30hp and a corresponding 30Nm of torque. That sees the 3.6-litre unit pull from lower revs with a muscularity that's missing with the standard engine. The effect of that is further enhanced by the owner specifying the fitment of a 996 Cup specification final drive, which shortens all the gear ratios and makes for improved acceleration in every gear. That combination makes for ferocious pace at lower revs, to the point that more often than not I find myself short shifting at around 5,500rpm, the car feeling indecently brisk when doing so, certainly on the road.

Up the revs further and that increased performance keeps on coming. This GT3 RS loses none of its high-rev ability, despite what it's gained elsewhere. It feels a bit more rounded on the road and the modifications would be sensational on track, with that greater low- and mid-range bringing additional pace in slower, more technical corners, while losing none of the high-rev rush on the straights. There you'd work the brakes harder too, being Alcon steel discs, which isn't unusual in tracked GT3 RSs. The brake pedal is firm and full of feel, with those discs washing off the easily gained speed. It's a car that's unquestionably rawer and more intense than its unmodified alternative, but it achieves that without removing the purity and finesse that make these cars so appealing to drivers.

Again, and as ever, clambering out I'm thinking how I might ever be able to buy one, especially as the 996 GT3 RS is approaching 25 years old, which means it can be easily imported into the USA. That, combined with its rarity and historical significance, will undoubtedly push values higher than they presently are, which means I'll likely never have one in my garage. Yet as long as there are owners and magazine editors willing, it's a car I'll happy revisit, as it really is one of the greats, if not the greatest – whether standard or not. **911**

THANKS The car in our pictures is for sale via RPM Technik (**rpmtechnik.co.uk**). For more information call **+44 (0) 1296 663 824**.

996 GT3 RS MR | 51

Total 911 verdict
The 996 GT3 RS has for too long been overlooked as one of the all-time greats. This car taps into the real RS philosophy of taking the base car and improving it, to great effect.

LIKES
- Modifications build on the foundations of a brilliant 911 to make it even more focused and fun

DISLIKES
- Slight trade-off in comfort

| Ultimate 911 GT3 Collection

992 v 991
GT3 Touring

Porsche's new Touring legend takes on the original…
which is the ultimate flat back 911?

Written by **Lee Sibley** Photography by **Ali Cusick**

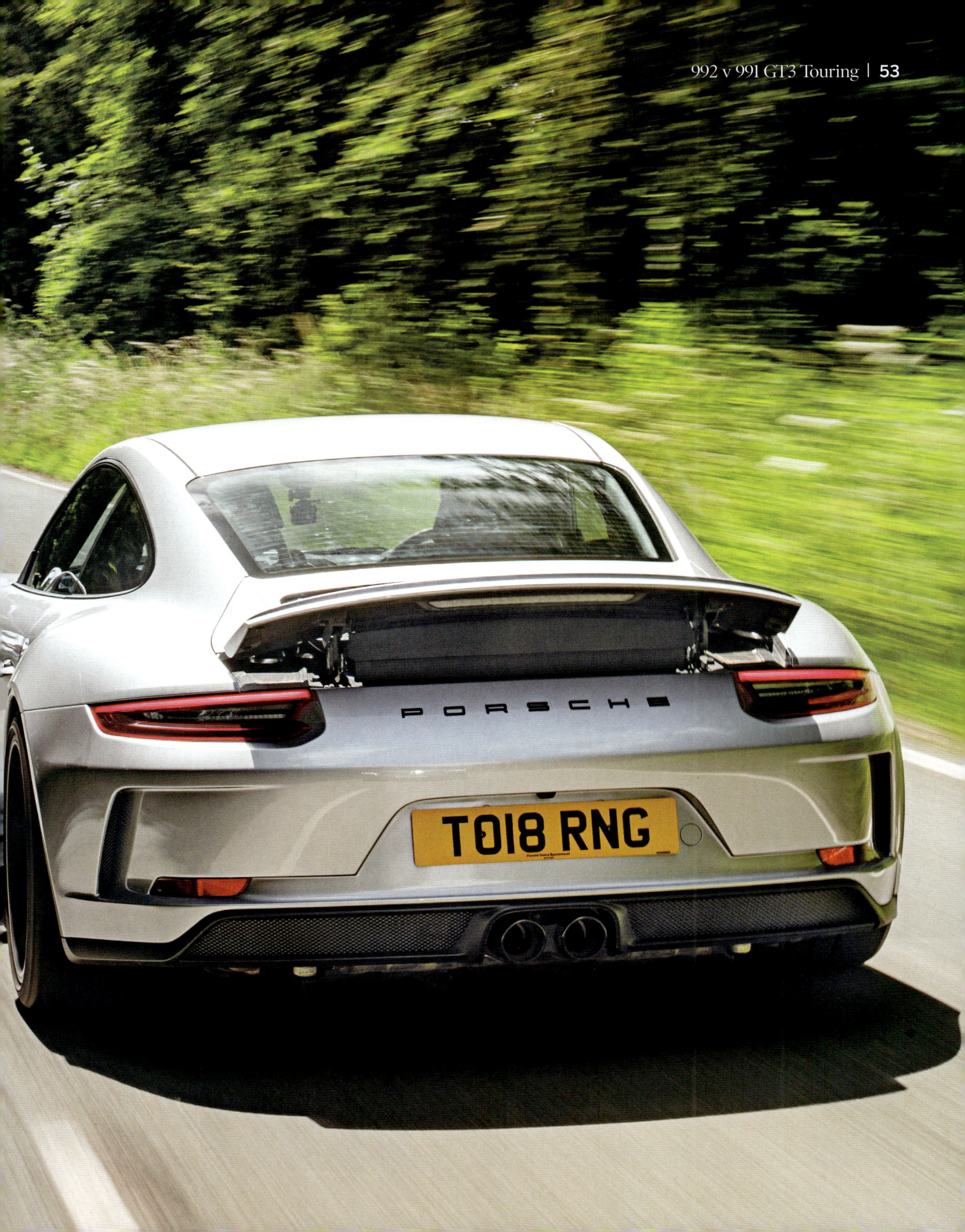
PORSCHE
TO18 RNG

54 | Ultimate 911 GT3 Collection

For 25 years and counting, Porsche's GT3 has cemented its legend within the 911 dynasty, one which enhances the 911's natural penchant for performance with additional flair and emotion. Powered by a high-revving, naturally aspirated flat six engine, fed to the rear wheels by – traditionally – a manual transmission, the GT3 is the purist's 911. The race version has accrued an envious horde of class wins in competition, and while the production version isn't as hardcore as the all-out Rennsport, itself very much a track car with licence plates, a GT3 nevertheless embodies the hallmarks of Porsche Motorsport.

However, since 2017, a new legend has been born out of the GT3 nomenclature. The Touring Package gave a new focus to the GT3's capabilities, its brief to refine that magic of the Motorsport department and make this 911 more suited to the road. The 991.2 GT3 was thus the first to bear the 'Touring' script on its decklid as part of a revised rear, which featured a traditional 911 'flat back' profile with active wing borrowed from its Carrera sister. Still retaining the winged GT3's explosive 500hp engine, the Touring's more subtle styling proved captivating.

That first Touring became an instant classic, and its star has never fallen, for several reasons. Seen as the everyday 911 R, even if key omissions such as a single mass flywheel ensured their specification is ultimately different, the Touring was offered as a no-cost option to those who'd specced their 991.2 GT3 with a manual transmission. Numbers therefore are low, with less than 50 to have reached UK shores.

We're not here to discuss collectibility, though. The 991 remains a scintillating drive, the devil very much in the detail over its fixed wing brother. The shift of its manual transmission is slightly smoother through the gate, the ride height has been adjusted and damping is slightly softer. The rest is pure GT in its noise, razor-sharp throttle response, and deftness at its nose. It's a mighty impressive package, the Touring more playful at road speeds without short changing you, should you want to turn the wick up on track. Andreas Preuninger and his GT department had nailed it: not since the 997 GTS of 2010 has a new 911 model line proved such an instant hit, and our refresher drive has underlined exactly why. How could Porsche improve on its GT3 Touring recipe?

On initial inspection of the new 992, it's clear it doesn't carry the 991's beauty through simplicity, with a clear uptick in aggression through styling. Its front and rear bumpers are more complicated in their appearance, though its adjustable wing on the back has a smoother profile without the embellishment of a Gurney flap as attached to the 991. The 992's 'nostrils' ahead of its front bootlid carry a clear synergy with Porsche's motorsport GT3s, but against the 991's more elegant, sculptural single vent, the 992's Touring looks a tad too fussy at the front for a road 911, and its markedly bloated appearance in general over its predecessor will split further opinion among purists – and as we've already established, this is very much a car for the purists.

Hold your judgement, because physics is very much behind the 992 Touring's form. Walking to its rear, I get on my knees and crouch down for a better view of its rear diffuser, which even when standing

---

Model 991 GT3 with Touring Package
Year 2018

Engine
Capacity 3,996cc
Compression ratio 13.3:1
Maximum power 500hp @ 8,250rpm
Maximum torque 460Nm @ 6,000rpm
Transmission 6-speed manual

Suspension
Front McPerson strut; some chassis bearings with ball joints; PASM
Rear Multi-link; some chassis bearings with ball joints; integrated helper springs; PASM

Wheels & tyres
Front 9x20-inch centre-locks; 245/35/ZR20
Rear 12x20-inch centre-locks; 305/30/ZR20

Dimensions
Length 4,562mm
Width 1,852mm
Weight 1,413kg

Performance
0-62mph 3.9s
Top speed 196mph

ABOVE 991 Touring was the GT department's first flat back 911, and was deemed an instant classic

992 v 991 GT3 Touring | 55

Ultimate 911 GT3 Collection

Model 992.1 GT3 with Touring Package
Year 2021

Engine
Capacity 3,996cc
Compression ratio 13.3:1
Maximum power 510PS @ 8,400rpm
Maximum torque 470Nm @ 6,100rpm
Transmission 7-speed Sport PDK

Suspension
Front Double wishbone with anti-roll bar; all chassis mounts with ball joints; integrated helper spring; PASM
Rear Multi-link; anti-roll bar; chassis bearings with partial ball joints; integrated helper spring; PASM

Wheels & tyres
Front 9.5x20-inch centre-locks; 255/35/ZR20
Rear 12x21-inch centre-locks; 315/30/ZR21

Dimensions
Length 4,573mm
Width 1,852mm
Weight 1,435kg (1,418kg manual)

Performance
0-62mph 3.4s (3.9s manual)
Top speed 197mph

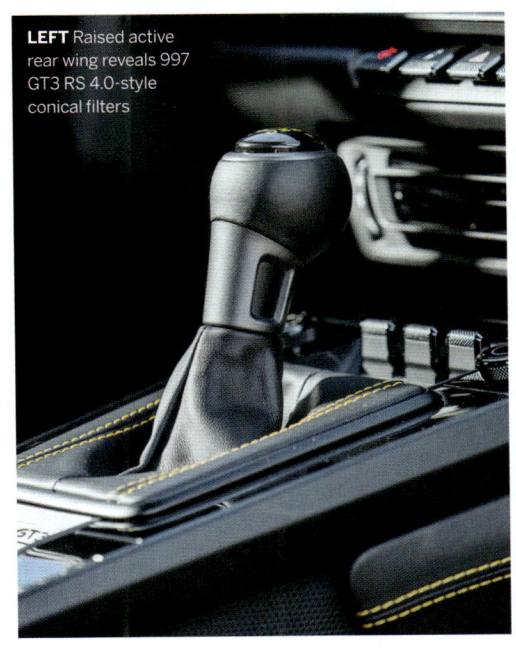

LEFT Raised active rear wing reveals 997 GT3 RS 4.0-style conical filters

up appears to have evolved significantly over the 991. Both wider and deeper, its longitudinal fins are fatter and travel further underneath this Porsche 911, making the 991's comparatively measly fins affixed to the base of the engine sump look like something of an afterthought.

Above that, and under the 992's active rear wing, is possibly my favourite detail on this new Touring, and one which flies in the face of long-running criticism that on both the 991 and 992-generation 911s, the iconic flat six engine is completely hidden. Not here: activate the rear wing from its resting position and in its place you'll see two air channels flowing down into two large inlets, each housing an exposed conical air filter. Alright, there's no flat six to gaze at, but the Touring's air filter setup is reminiscent of the 997 GT3 RS 4.0 which, by coincidence, is the last time we could readily observe a flat six housed in the business end of a 911. It's a beautiful detail, and well engineered (the active wing, when deployed, has a curtain running right along its leading edge, enhancing a ram air effect down into those conical filters). Enough at the looking, though. It's now time to climb inside the 992.1 Touring and head out onto the road.

If it wasn't for a tacho running all the way round to nine grand, you'd be forgiven for thinking you're taking a seat in a regular Carrera: much like the 991, there's no Alcantara like in the regular GT3, the wheel and shifter covered instead by rich leather as part of the Touring's unique interior package. The 12 o'clock marker on the wheel, usually highlighted in yellow, has been swapped out for more black leather, while the 918 buckets are clad in leather with durable RaceTex centres. There's no cage behind, though there aren't any rear seats either.

About the shifter: short, stubby and spherical, you could well expect to find a 6-speed H-pattern emblazoned on the top – and in a manual Touring you most certainly would, except this isn't a manual Touring. For the first time, Porsche is offering its Touring with PDK, and so my selection choices are P, R, N or D, with the shifter only afforded a forward or backwards motion. It's an intriguing development for a 911 designed in the spirit of the R.

Unperturbed, I start the engine, the theatre of which is identical to the 991 in its crank-over and eventual fire, its throaty rumble permeating through the GT3's cabin. Nudging the shifter down to 'D', I head out into the green countryside surrounding Sussex's Goodwood estate.

It takes a matter of minutes, and only one corner, for my jaw to hit the floor. The 992 Touring's front axle is the source of my amazement: never before have I experienced such a crisp, responsive front axle on a Porsche 911. So much so I find myself having to recalibrate the timing of my inputs at the wheel to avoid diving in for the apex prematurely, so quick is the nose to follow inputs at the wheel. The 991 Touring is no slouch in this regard, yet compared to the 992 its system feels slightly lethargic to lumps, bumps, cambers and corners. And I never, ever thought I'd be writing that about a 991.2 GT3.

The steering – oh the steering! – meanwhile is alive with feel in the 992, its wheel chattering away in the palms of my hands. It's a sensation not found in the 991, which is a little more stoic in the way it goes about its business, filtering out a lot of noise from the road's surface, particularly at its centre point. Again, the 991's system is highly commendable, a class leader in the age of Porsche's electrically assisted steering, but the fact is in the 992, every last slither of information from the road's surface is being fed back to my hands. That doesn't mean I'm wrestling to keep this 911 in a straight line over bumpy surfaces (and there are many here) like I would in say a 996 or 997 GT3 RS, the electric assistance doing its job here, albeit not to the point of what feels like undue intrusion. Quantifiable as a '991.2 RS Plus' setup, not since the days of hydraulic assistance has the 911's steering feel and communication been so complete.

There appears to be more mechanical grip too: the 992 Touring is just as playful on the road as the 991, but on previous occasions in the latter I've been caught out, its rear end stepping out quickly and seemingly without warning as the car attempts to keep 500hp and 460Nm connected to the floor ➔

58 | Ultimate 911 GT3 Collection

with little aerodynamic aid. The 992 benefits from an extra half-inch of wheel connecting it to the road at each corner, while its active rear wing, which appears to deploy to its upright position more often in Sport mode, does a fine job of helping to squash the rear of the 911 into the road.

The 992 Touring is stiffer than the 991, the latter of which displays a little more body roll on the same corners. The 992 is not overly stiff though, despite those ball joints front and rear, an integrated helper spring adding a layer of compliance to the ride. It's another high point to the 992's arsenal, its chassis somehow taking those juxtaposed ideals of agility and comfort and admirably melding them together.

All the while, the 992's 4.0-litre flat six is serving its usual GT3 sensations: there's an extra 10hp and 10Nm over the 991, with individual throttle valves and more efficient injectors taken from the 991 Speedster, the performance on offer being truly explosive. Power, delivered instantly via a razor-sharp throttle, is accessible in abundance from any point in the rev range, the rush taking you wilfully right up to that 9,000rpm redline – if you leave your right foot in – with minimal drop-off at the very top end.

Noise here is most interesting too, in that the GT3's now customary shriek isn't quite as guttural between eight and nine grand compared to the 991. You can thank the 992's gasoline particulate filters for that, the GPFs deployed on all post 2018 MY Porsche 911s in order to comply with Euro 6 emissions regulations. That the 992's soundtrack is recognisable and (in isolation) evocative to the GT3 signature is nothing short of miraculous. On the turbocharged Carrera cars, those GPFs really do thwart the 911's flat six symphony. Here though, in a tangible triumph over adversity, Andreas Preuninger's GT department has done a splendid job of allowing its new GT3 to sing – only in the company of the 991 does a slight deviation manifest at the very top end.

Staying with sound, the 992 does a splendid job of muting a lot of droning inside the cabin which adversely impacted the 991 Touring's experience. It was perhaps the great criticism of the original, and even led to some 991 owners adding more sound insulation back into the car to combat excessive tyre and exhaust humming over long journeys, which was at odds with the Touring name. To combat this, the 992's windows are bi-planed, filtering out a lot of the humdrum. Aurally then, it's a case of more of what you want, and less of what you don't – a huge win for the 992 Touring in terms of its all-round appeal.

So the chassis, soundtrack and flat six is stupendous on the 992 GT3 Touring… but does that PDK transmission make or break it? Emphatically, it's the former. This might be controversial, but I believe PDK is better suited to the Motorsport department's 4.0-litre flat six over the 6-speed manual equivalent, its performance capabilities on par with that brilliant boxer. On a squiggly mountain pass, PDK will still score high for engagement in the latest Touring despite only having two pedals, allowing for fast gear changes up and down the 'box to exploit the 4.0-litre's impressive power band in the top half of the tacho. As we know, another part of the touring adventure is actually reaching your favourite ribbon of road, and here again PDK is the answer, delivering both comfort and efficiency that a manual can't match. Porsche purists will baulk, but surely PDK actually feeds in to the mantra of what a genuine tourer is all about?

There are downsides to the 992.1 Touring though, the chief gripe being its size. On narrow, British B-roads, the 992 feels absolutely huge, and placing it on the asphalt – especially here in left-hand drive – isn't exactly straightforward. It's mostly a placebo effect, because the 992 Touring isn't any wider than the 991 Touring, both measuring 1,852mm across their rumps at peak width. However, the 992.1 Touring's front axle is much wider, even to the eye, given away by the 992's more bulbous front arches over the 991's comparatively slender fenders. The fatter body accommodates an increased front track of some 48mm over the 991, and you don't half feel it behind the wheel. It's great for road holding and flatter cornering, but not so great when attempting to negotiate tight country roads.

The 991 Touring might also win more enthusiasts over with its prettier looks, but as a tool for the joy of driving, there's no question the 992.1 Touring has moved the game on. This road-biased recalibration of the modern GT3 is a fine follow-up to that revered original, and stakes a real claim to the title of quintessential Porsche Q-car. The ultimate flat back 911? A 911 R might have something to say about that, but a Porsche 911 GT3 Touring is simply the more attainable solution for most of us… 911

ical Ultimate 911 GT3 Collection

BATTLE
— OF THE —
BEST

The 991 GT3 RS is a technological phenomenon and a worthy Rennsport for the digital age. But for all its precision, can it match the analogue thrills of Andreas Preuninger's finest? We find out on track and road…

Written by **Josh Barnett & Lee Sibley** Photography by **Ali Cusick**

991 GT3 RS vs rivals | 61

At its most reductive, the idea that certain activities can 'make you feel alive' is a peculiar one, especially when you consider the flipside: I have certainly never done anything that has made me feel dead. Yet this supposedly tangential notion is never more evident to me than when I am out on a racetrack, pushing a car to its limits. The often delicate and sometimes brutal dance on the edge of adhesion from corner to corner is enough to get thousands of petrolheads' pulses racing. It is a sensation that is intrinsically woven into the fabric at Zuffenhausen and it is, therefore, the key ingredient in what is undoubtedly the 911's most exciting and renowned subdivision: Rennsport.

Based near the race teams in Weissach, Andreas Preuninger's GT cars department are the current custodians of this legendary moniker. This crack squad of engineers has proven that they truly understand what is needed to create an enthralling Neunelfer experience, with a track-focused character that is equally captivating out on the open road. Nowhere is this more apparent than in the 997 generation of GT3 RSs. From the 3.6-litre, first-generation iteration to the instantly iconic 997 GT3 RS 4.0, Preuninger's team never missed a beat between 2006 and 2010, somehow managing to improve on perfection with each revision, culminating in the aforementioned 4.0-litre Rennsport – a car that we concluded in issue 125 was "the king of kings". These days, the RS ranks have been bolstered further with another 3,996cc pretender to the RS 4.0's throne.

The 991 GT3 RS is, on paper, the antithesis of the 997's analogue thrills: a PDK gearbox in place of the lauded six-speed manual shifter, a flat six based (loosely) on the Carrera's 9A1 engine rather than the motorsport-derived Mezger, and rear-wheel steering in place of the previously passive back axle. These changes have made the 991.1 RS devastatingly effective and hugely coveted, just like its 4.0-litre 997 forebear.

That was in isolation though; context is key here, which is why we have gathered both 4.0-litre Rennsports (as well as both previous generations of the 997 GT3 RS) together for the ultimate test on track and road. As a supposed standard production model, the 991 is intended to be the successor to the 3.8-litre 997.2 GT3 RS. However, I'm going to start with the RS 4.0. After all, to paraphrase De La Soul, "four is the magic number", especially in the world of water-cooled Porsches.

When it was released in 2010, I couldn't believe that the 997 GT3 RS 4.0 was road legal. More so than any Rennsport before it, it looked like a race-ready 911. Those dive planes and that rear wing (taken straight from the 997 GT3 Cup car) have never failed to catch my attention. Yet, sat alongside its successor, my gaze is very quickly diverted towards the 991. Mounted higher than ever before, the rear wing is even more of a focal point on the latest RS and, combined with those front arch louvres and induction scoops on the Turbo-width flanks, the 991 GT3 RS doesn't need garish decals to capture my attention. It makes the RS 4.0 look ordinary.

From behind the wheel, however, the 997 GT3 RS 4.0 certainly doesn't prove itself to be ordinary. In fact, on both road and track, it is anything but. The ➲

Ultimate 911 GT3 Collection

Track test:
991.1 GT3 RS vs 997 GT3 RS 4.0

997.1 GT3 RS: LEFT BEHIND?

It's hard to believe the first-generation 997 GT3 RS is now over 17 years old, but when you digest that time as over 7,000 days of engineering evolution having elapsed at Weissach, you can be forgiven for dismissing the early 997 Rennsport's technology as largely dated. The first track 911 to get PASM as standard (but not dynamic or active engine mounts), it is the only Rennsport of our quartet on test not to arrive with that coveted 'five star' Total 911 rating. Have we been harsh?

In issue 135, we climbed behind the wheel of this first 997 RS in isolation from its younger Rennsport brethren, where our original 4.5-star rating was found to be justified. We said, "With a heavier flywheel than the 997.2 and 35bhp less power, the 997.1 feels a less aggressive package. I'm not as on edge behind the wheel as I want to be." We then concluded, "With Gen1 cars retailing for less than 10 per cent under the price of a Gen2 Rennsport, you'd be mad not to stump up that little bit extra required for the keys to a 997.2 example."

On reflection, the 997.1 GT3 RS has always endured something of a tumultuous reputation. Even from release, commentators pointed to the fact it shared the same performance figures as its GT3 sister (again, the only RS here at our Silverstone test to do so), shedding just 20 kilograms of weight in the process. Even the Porsche crest on its bootlid was a point of contention: merely a sticker on the 996 GT3 RS in homage to its motorsporting credentials, Porsche reverted back to a heavier metal emblem for the first 997. The real-world difference may have been a matter of grams but there was a principle to enthusiasts' outcries.

Of course, the first-generation 997 GT3 RS's time at the top of the 911 performance tree was short lived, replaced only two years later by the second-generation, 3.8-litre Rennsport (the first time two or more Rennsports have been contrived in the same generation of 911 since the 964 some 15 years earlier). The 3.8-litre car improved suitably on the shortcomings of the 3.6-litre variant and ever since then the predecessor has rightly lived in the shadow of the successor. And, against today's 991, the 997.1 is very nearly a whole second slower to 62mph, a relatively huge gap in what is but an incremental measure of a car's performance.

Despite this, the 997.1 GT3 RS is still a superb 911, boasting feedback and weighting at the wheel that the electrically assisted 991 can only dream of. In fact, when all is said and done, the 997.1 very much delivers that Rennsport spirit craved by so many – it's a shame that three of its contemporaries are just so much better.

driving experience of this limited-edition Neunelfer is fittingly defined by the flat six powerplant from which its name is derived. Closely related to the similarly sized engine in the 997 GT3 Cup and 997 GT3 R racers, the Mezger in the RS 4.0 is ripsnorting proof that you can really have your cake and eat it. Compared to the 3.8-litre unit in the 997.2 RS, the extra low down torque helps you to punch out of corners with impressive verve, yet this is not an engine that solely thrives in the mid-range. Letting the Mezger run out all the way to its 8,250rpm redline brings a symphony of aural pleasures that combine at the top end to produce a hair-raising mechanical melody. It's absolutely addictive.

The 991.1 GT3 RS's 9A1 engine provides a very similar dynamic character, with the 4.0-litre architecture providing the shove that is lacking in the 991 GT3. However, there's something missing in the 991 RS's soundtrack. Where the standard GT3 finishes with its banshee-like 9,000rpm flourish, the 8,800rpm-limited RS lacks that final crescendo. What's more, while there's a pleasant organic-ness to the RS 4.0's note, the 991 sounds too... perfect. It's too refined and sounds too much like a steroidal Carrera to get my pulse truly racing.

Where the 991 really excels, though, is its chassis. While the RS 4.0's steering feel is undeniably more intuitive (which is the result of the hydraulic power assistance rather than the 991's EPAS), the 997 is indeed still hampered by the idiosyncratic Neunelfer flaws. With all that mass over the rear end, the RS 4.0 is more prone to understeer on corner entry although, despite the canards, the front-end aerodynamics are still overpowered by the huge rear wing, causing the steering to go light, especially in medium-speed corners.

Understeer isn't even a concern in the 991, though. Its turn-in is so direct that it's almost un-911-like. It's become fashionable to attribute the latest Rennsport's nimbleness to the rear-wheel steering system, but on track and during fast road driving it is more likely to lengthen the wheelbase than shorten it. Instead, the wider front track of the 991 enables a softer front anti-roll bar without compromising roll control, providing the front end with more bite through each corner. Coupled with more mass on the nose (the result of moving the engine forward on the 991 platform), it means that the new GT3 RS is a much less compromised track tool. Although it never feels like a car reliant on downforce, I'm sure those eye-catching aerodynamic devices help the overall grip levels too, especially on circuit. After all, this is a car capable of 1.7G lateral loads, on road-legal tyres.

The caveat with the 991, though, is that, at the limit, it is more likely to suddenly bite you than the RS 4.0. While the 997's steering and chassis is more progressive, the 991 doesn't telegram its dynamic messages to you as effectively, creating a snappiness that makes it less approachable to Rennsport rookies.

Despite its talents on track, the 991 seems equally at home on the road, too. Unlike previous RSs, the latest iteration's damping makes the car feel beautifully pliant over the bumpy British back roads that such a car should really thrive on. Combined with a steering system that filters out some of the harshest cambers around the centre point, it leaves you to enjoy pinning the 991 to each apex with prodigious pace and accuracy. If I had one complaint, it would be that, like the 9A1 engine under its decklid, the 991 often feels too refined. It lacks the raw emotion so often associated with those other iterations that are lucky enough to wear the Rennsport badge.

I certainly can't say the same of the 997. If anything, thanks to the rose joints on the rear suspension, the RS 4.0 feels too fidgety on the open road. Despite this, with that delectable manual gearbox, the 997 is unquestionably the more involving experience. Combined with that delightfully communicative steering, the RS 4.0 is the last Rennsport built to satisfy Porsche 'purists'. The 991 may be the undoubted king of the racetrack, but it doesn't have exclusive rights to the RS crown. The two 4.0-litre legends will have to learn to share.

While the 997 GT3 RS 4.0 more often than not finds itself more at home on track rather than road, ➲

Ultimate 911 GT3 Collection

Model	991.1 GT3 RS (2015)	997.2 GT3 RS (2009-12)	997.1 GT3 RS (2006-07)	997 GT3 RS 4.0 (2010)
Engine				
Capacity	3,996cc	3,797cc	3,600cc	3,996cc
Compression ratio	12.9:1	12.2:1	12.0:1	12.6:1
Maximum power	500hp @ 8,250rpm	450bhp @ 7,900rpm	415bhp @ 7,600rpm	500hp @ 8,250rpm
Maximum torque	460Nm @ 6,250rpm	430Nm @ 6,750rpm	405Nm @ 5,500rpm	460Nm @ 5,750rpm
Transmission	Seven-speed PDK automated manual	Six-speed manual	Six-speed manual	Six-speed manual
Suspension				
Front	Independent; MacPherson strut; PASM dampers; coil springs; anti-roll bar	Independent; MacPherson strut; telescopic dampers with coil springs; anti-roll bar; PASM	MacPherson strut; coil springs; anti-roll bar	Independent; MacPherson strut; telescopic dampers with coil springs; anti-roll bar; PASM
Rear	Independent; multi-link; PASM dampers; coil springs; anti-roll bar	Independent; multi-link; telescopic dampers with coil springs; anti-roll bar; PASM	Multi-link with telescopic dampers; coil springs; anti-roll bar	Independent; multi-link; telescopic dampers with coil springs; anti-roll bar; PASM
Wheels & tyres				
Front	9.5x20-inch centre-locks; 265/35/ZR20 tyres	9x19-inch centre-locks; 245/35/ZR19 tyres	8.5x19-inch alloys; 235/35/R19 tyres	9x19-inch centre-locks; 245/35/ZR19 tyres
Rear	12.5x21-inch centre-locks; 325/30/ZR21 tyres	12x19-inch centre-locks; 325/30/ZR19 tyres	12x19-inch alloys; 305/30/R19 tyres	12x19-inch centre-locks; 325/30/ZR19 tyres
Brakes				
Front	380mm discs with six-piston calipers	380mm discs with six-piston calipers	380mm discs with six-piston calipers	380mm discs with six-piston calipers
Rear	380mm discs with four-piston calipers	380mm discs with four-piston calipers	360mm discs with four-piston calipers	380mm discs with four-piston calipers
Dimensions				
Length	4,545mm	4,460mm	4,460mm	4,460mm
Width	1,880mm	1,852mm	1,808mm	1,852mm
Weight	1,420kg	1,370kg	1,375kg	1,360kg
Performance				
0-62mph	3.3 secs	4.0 secs	4.2 secs	3.9 secs
Top speed	193mph	192mph	194mph	193mph

997.2 GT3 RS 3.8
- Better front end grip than 997.1
- Much-improved levels of downforce
- Hugely undervalued in current market
- Power delivery is sluggish below 4,000rpm
- RS 4.0 shows weight could easily have been further reduced from the factory

997.1 GT3 RS 3.6
- Rennsport package in 997 specification is sublime
- Steering weight and feedback is better than 991
- Unrefined aero means car is fidgety at high speed
- No power increase over requisite GT3
- Lacks aggressive visual appeal of its successors

991.1 GT3 RS
- Cup-rivalling performance in a road car
- PDK Sport is supremely intelligent, smooth and lightning quick
- Overall chassis balance is the best ever in a 911
- Simply too fast and precise to enjoy on public roads
- DK not as involving as a third pedal
- Pit speed limiter is pure gimmick

997 GT3 RS 4.0
- Increased torque at lower revs allows for cornering in higher gears over 3.8-litre 997 Rennsport
- This is the Mezger engine in its final, most glorious form
- Stiff chassis and improved aero
- Passive axle lacks poise on corner entry over 991
- Collector appeal means many examples are never likely to see a track

Road test:
991.1 GT3 RS vs 997 GT3 RS 3.8

the 3.8-litre 997 Rennsport can be considered as 'the everyman's RS'. Yes, it's still a £175,000 Porsche 911 but, in such accomplished company, such terms are all relative. It's less of a collector's piece than the 4.0-litre car and, as such, is more likely to find itself used as Preuninger's team intended. What's more, as the last full production RS, the second-generation 997 is actually the true predecessor to the 991; with the 4.0-litre link it just seemed rude not to invite the RS 4.0 along first.

Sliding into the 997.2 GT3 RS, it instantly feels like a truly purposeful place to perch yourself. While 918-style seats in the 991 GT3 RS provide excellent support, there's real drama as I shoehorn myself into the Nomex-clad Recaro bucket seat in the 997, while the removal of the air-con and PCM units in this particular car makes it clear what this Rennsport's intentions are before I've even turned the engine over. By comparison, despite the new steering wheel in the 991 RS, the cockpit feels like a more generic environment (although if the centre console went on a similar diet to the 997 I'm sure it would feel at least a little more special).

Starting the 997 is a similarly characterful experience, as that legendary Mezger fires into life with a snarl, settling into an angry, recalcitrant idle. The throttle pedal has an immediacy that causes the 3.8-litre unit to bark gregariously with a single,

sharp prod; like all the best Rennsports, the 997.2 is a highly strung thoroughbred. It's a flat six that loves to live in the upper echelons of the rev counter, feeling relatively dead below at least 5,000rpm – there isn't the same punchy mid-range torque as found in the RS 4.0. This isn't a bad thing, though, as even on the open road the 3.8 RS's peaky nature encourages me to let it off the leash. Beyond 6,000rpm, the 997.2 really begins to take off, supported by a gloriously mechanical growl that rewards you for chasing the redline through every single gear.

As the shift light blinks on just beyond 8,000rpm, I lift briefly, snapping from second to third before getting back on the throttle to do it all again. Each gear change is met with an intoxicating machine gun-like chatter as the Mezger refreshes itself, ready for another run towards the horizon. The whole symphony is backed by the induction hiss as the engine greedily sucks in more sustenance. There is only one modern 911 that sounds this good: the 991.1 and 991.2 GT3.

Without the rose-jointed rear suspension found on the RS 4.0, the 997.2 GT3 RS makes for a superb B-road blaster. Despite the 997 chassis' flaws, the hydraulic power steering system always lets you know what the front end is doing and, on the public highway at least, understeer is very rarely a real problem. Let there be no mistake, compared to most

911s, the 997.2 RS enjoys prodigious amounts of grip (even on damp tarmac and Michelin Cup 1 tyres). However, it doesn't have too much adhesion. Unlike the latest batch of Neunelfers, I'm very much the key component when driving the 997 RS and I'm having to concentrate completely to keep up with Lee (who is setting an impressively rapid pace in the 991). It makes for an addictive experience as I delicately balance the 997 through a succession of sweeping bends and, when it all goes perfectly, the whole thing is hugely rewarding.

Jumping out, I'm sweating a little (though that may just be the lack of air-con) and my arms have evidently had a work out as the 997 hunts around on cambers and bumps, but I just can't stop smiling. The 991 GT3 RS has some big boots to fill emotionally. It starts well, firing up with a convincing impression of previous Rennsport 911s. Those imposing air intakes on the rear arches really help to amplify the induction sound, too; if you thought they were there just for show, put your hand over one and prod the loud pedal. This may be an RS for the digital age but it seems to still have the 'show' as well as the 'go'.

Compared to the 997, the 991's extra capacity and improved induction definitely bring more thrust around the lower reaches of the rev counter: it's become a huge buzz word in automotive marketing circles but the latest RS is infinitely more tractable.

| Ultimate 911 GT3 Collection

Gearbox
PDK: 991 GT3 RS
MANUAL: 997 GT3 RS 4.0, 997.2 GT3 RS, 997.1 GT3 RS

MAX RPM
- 991 GT3 RS: 8,800rpm
- 997 GT3 RS 4.0: 8,500rpm
- 997.2 GT3 RS: 8,500rpm
- 997.1 GT3 RS: 8,400rpm

CYLINDER CAPACITY
- 991 GT3 RS: 666
- 997 GT3 RS 4.0: 666
- 997.2 GT3 RS: 633
- 997.1 GT3 RS: 600

Nürburgring Lap times
- 991 GT3 RS: 7:20
- 997 GT3 RS 4.0: 7:27
- 997.2 GT3 RS: 7:33
- 997.1 GT3 RS: UNRECORDED

POWER TO WEIGHT
- 991 GT3 RS: 352 HP/TON
- 997 GT3 RS 4.0: 368 HP/TON
- 997.2 GT3 RS: 328 HP/TON
- 997.1 GT3 RS: 302 HP/TON

991 GT3 RS vs rivals | 67

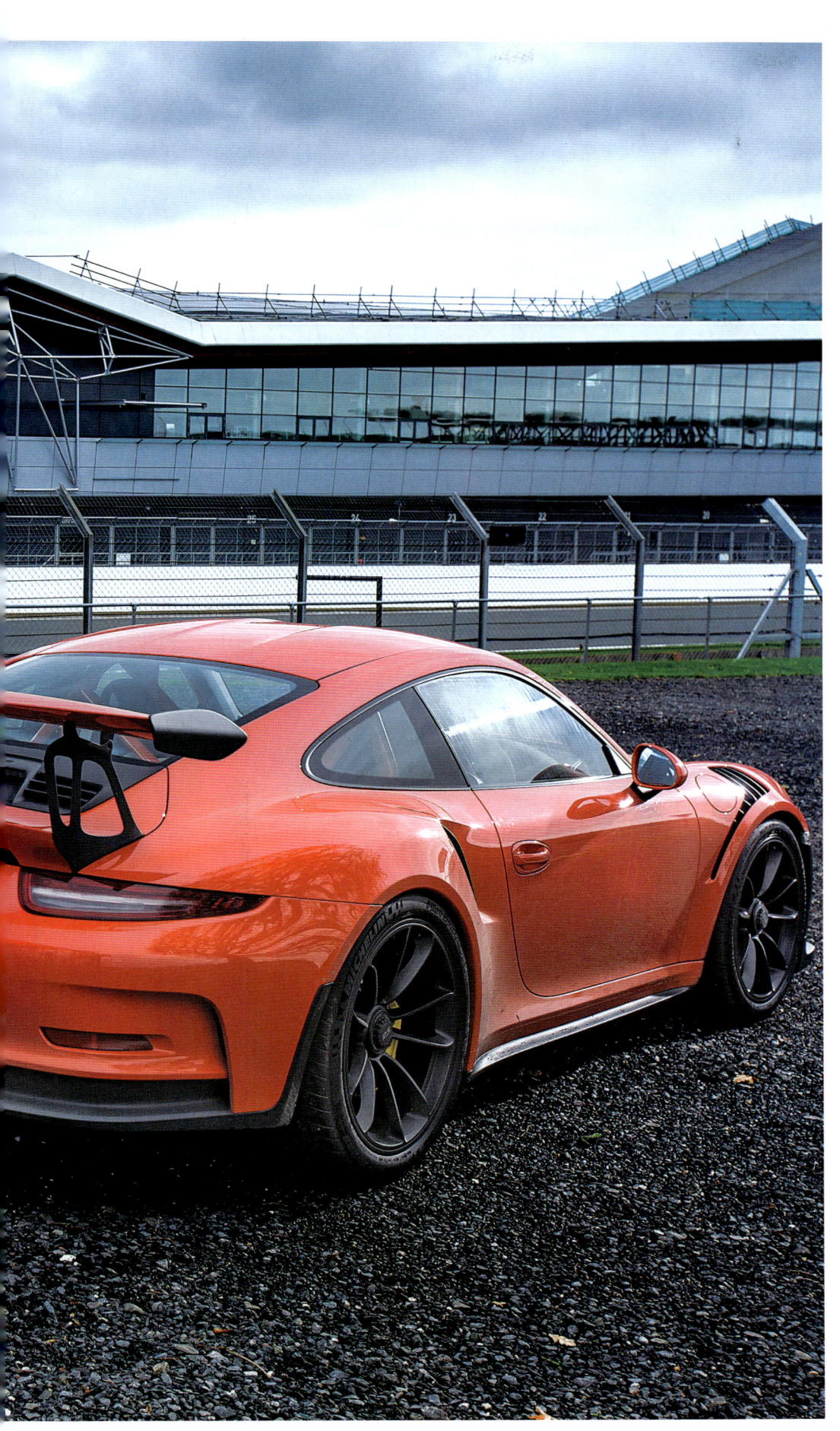

It does mean that there is less incentive to wring the neck of the 991.1 GT3 RS as I'm not required to head for the limiter to make progress. There's no real aural reward at the upper end of the rev range either, with a clinically aggressive sound throughout each sweep of the needle. Compared to previous Rennsports (and subsequent GT3s), the last few hundred rpm are something of an anti-climax. If anything, the 9A1 in the RS feels like it is running out of puff more keenly after 8,500rpm than the similar unit found in the GT3.

This is not to say it feels slow. Far from it. There's an effortless pace to the latest Rennsport and, even with less mass over its rear wheels than the 997, it's able to put its power down more effortlessly, too, thanks to those huge 325-section rear tyres. Where the 997.2 is spinning up in second and third gear, the 991 is instantly planted, shooting forward with greater verve and inspiring more confidence mid-corner, vital in the damp and wet conditions that we're often blessed with here in the UK. The damping feels slightly softer, too, meaning that bumps are less likely to upset the 991's balance. The 991 gives me much greater confidence from the chassis on turn-in, too, darting its way towards each apex with minimal fuss, while mid-corner adjustments are possible now too. On rare occasions when it doesn't want to play ball, you can simply trail brake into the turn, too, a benefit of the switch to the two-pedal PDK shift setup.

The gearbox feels even faster than the standard GT3, with each change dispatched with a violently efficient crack at the slightest touch of the weighty metal levers. The technological prowess of the system is mind-boggling, and it does make you feel like you're driving a real 911 GT3 R or RSR but, for all its ability on the track, on the road it does feel like some of the skill necessary for previous RSs has been taken out of my hands. The 997's delicious steering feel has disappeared in the transition to the new generation of electric systems, too. While the 991 RS's EPAS is by far the best I've driven in terms of communication, the messages supplied to my fingertips still feel vague in a direct back-to-back with the 997, and the weighting is, in comparison, too artificial.

Ultimately, this is the crux of the issue with the latest Rennsport. As a piece of engineering, it is simply unrivalled; I can't doff my cap enough towards the GT cars department at Weissach. With every mile that I drive in the 991 GT3 RS it continually astounds me with its prowess, but there's a little bit of me that is left cold by the car's clinical ability to counter all that faces it. On the road, it's simply too able for its own good.

The thing is, on the track, the 991 is mighty, its full technological repertoire coming to the fore. In fact, there is so much grip that it sometimes seems like the laws of physics are being wilfully broken, while the 997's dynamics – which made it so endearing on the road – make it feel like you're always battling a compromised package. Therefore, my only logical conclusion is that you really need both. Yes, seriously, both. In their own ways, they highlight the very best of what Porsche can achieve: the 991 is the blue-sky thinking side of those in Stuttgart, while the 997 is redoubtable heart and soul. *911*

Ultimate 911 GT3 Collection

991 vs 992 GT3
BATTLE OF THE GEARBOX

Both these Weissach superstars are equipped with the same six-speed manual gearbox from Porsche Motorsport... but can Total 911 find any tangible differences between the two?

Written by **Lee Sibley**

991 vs 992 GT3: battle of the gearbox | 69

North Wales. A glorious pocket of the UK, where the locals are friendly, the scenery is pretty and the roads are twisty. Never mind God's country, today it's GT3 country, our two stunning examples from the 991 and 992 generation playing cat and mouse on the slithers of blacktop running through vibrant, green topography.

Ahead of me, Phil Farrell is carving a fast line through the cambered curves in his Carmine red 991.2 GT3, the howl from its exhaust reverberating around the Welsh hills. I'm in hot pursuit in a Gentian blue, Touring specification 992 GT3, its low, wide nose hunting the tall uprights and shark-fin end plates mounted to the 991.2's sculptural wing. We've covered these two glittering Porsche 911s in broader detail before, but today we're back on winding asphalt to delve a little deeper into just one aspect that makes them both so special: their gearboxes.

The history of Porsche's six-speed 'box in the company's GT product is well known. Binned for the 991.1 GT3 generation in 2013 when Andreas Preuninger famously announced PDK would be the de-facto transmission for Porsche GT cars going forward, a public outcry led to this decision being reversed. It was the 911 R in 2016 that first revived a manual stick-shift in a Weissach car, the U-turn complete when Porsche introduced the 991.2 GT3 a year later, when enthusiasts could choose either PDK or manual. By 2018, Porsche was producing a Touring version of said GT3, where any choice of transmission was once again removed, this time with stick shift being the compulsory gearbox.

For the current 992 generation, both manual and PDK are once again on the table, this time for both winged and Touring variants. Thus, equilibrium has returned to the Porsche GT stable, certainly in the eyes of purists who believe a manual transmission is to be at the epicentre of a car delivering such a visceral, driver-focused experience.

And so to today, and the B4391 in Snowdonia. We've sung the praises of this road in **Total 911** many times before, the 8.2 miles from Ffestiniog towards Bala being a road we know very well. It's the perfect stage to get these two exquisite GT3s dancing, and means I can focus more on the intricacies of their gearboxes rather than whether the next corner is likely to be a long left or hard right.

Both these GT3s use a six-speed manual gearbox to put 500-odd horses through the rear wheels (via a mechanical LSD), developed by the 'wizards of Weissach' at Porsche Motorsport. In fact, these two manual transmissions in the 991 and 992 GT3 ➲

BELOW While the shifters in both cars are just as short as each other, it requires less movement in the 991 to change gears

991 GT3

share rather a lot… because on paper they're exactly the same.

That's right: the same number of gears, identical ratios (*see page 46*) and even mass – although while we can't ascertain the exact weight of Porsche Motorsport's six-speed manual gearbox (unless anybody has a GT3, a set of scales and a spare afternoon?), we do know a 991 GT3 is 17kg lighter if it has a manual gearbox fitted rather than a PDK. This 17kg difference is also evident in manual-to-PDK 992s, even though the 992 platform overall is 5kg heavier than the 991.

So if both gearboxes are the same, what on earth are we doing on a chilly spring morning in Wales, trying to split hairs?

Well, dear reader, you should know by now that with Porsche, identical numbers on paper don't always correlate to the same physical feeling from where it matters – the driver's seat. Just ask anyone who's driven a 381bhp 996.2 GT3, and a 381bhp 996 GT3 RS. Porsche has proven time and again over the years that these cars really are more than the sum of their parts, and so splitting hairs is all part of the fun on this magazine.

With the GT3s still warm from our early-morning game of cat-and-mouse, I slide into the driver's seat of the 991. Turning the engine over, it settles to a bassy, burbling idle, offset by a metallic chattering of the single mass flywheel. And right there, we've already stumbled across a key difference between the two cars: the 991 being winged means it comes with an optional single-mass flywheel (this isn't available on Touring-spec cars in the 991 and 992 generation, to help differentiate between them and that halo 911 R).

Pulling on to the road and again chasing Phil, who this time is piloting the 992 Touring, it takes all of a couple of hundred yards – and two cog swaps – to realise there's very much a difference between the shift in the 991 and 992 GT3.

For a start, it feels as though there's less travel to the shift in the 991 over the 992. The short, stubby form of both shifters is identical, and so the sensation isn't synthetic, either. In the 991 there's simply less movement required to move the shifter into each gear. It's not quite as rifle-bolt as the legendary 911 R, but it does give a tighter, more direct feel over the 992. It's marginal, don't get me wrong, but very much apparent.

There's more weight to the 991 GT3's shifter itself too, with a definitive snick through each gate, reminiscent of the excellent, feelsome throw in a 997.2 GT3 RS. It all adds up to a more mechanical feel in the 991.2 GT3 when compared to the 992, the shifter of which feels superficially light at times. Again, the margins are slight, but clearly identifiable.

The 992's shift by contrast is more fluid, and you can really throw its lever around with ease. It's ideal for tighter sections of these twisting roads where cog swaps are more frequent.

What's most interesting here is that, if anything, you'd expect this difference to be the other way around. After all, the 991 GT3 with nearly 10,000 miles on the clock has been 'worn in', whereas the 992 has only just surpassed its 932-mile breaking-in period. Nevertheless, it can be argued these idiosyncrasies play rather nicely into the specific functions of these two GT3s. The winged car, being more track focused, requires an altogether more aggressive driving style for fast lap times, and so a little more weight behind the shifter may aid a more positive throw. The Touring, in contrast, is designed to deliver high engagement

BELOW The 992's shifter is lighter in the hand than the 991's, which suits multiple gear changes along a twisting section of road

992 GT3

"I myself prefer the weightiness of the *shifter* in the 991, yet find favour with the fluidity of the *shift* in the 992"

on winding mountain roads, where a fluid shift could be more welcome.

The clutch pedal on the 991 is lighter than that of the 992 – which you'd expect, because the 991 is rocking the single-mass flywheel compared to the 992's dual-mass item. However, the difference, certainly in terms of feel, isn't as great as you'd expect.

Both cars and their gearboxes come equipped with an arsenal of tech to help the driver get the most from them. An example of this is a rev-matching feature. Purists may baulk at its very concept, let alone presence, yet it actually works very well. Engaged automatically in Sport mode, the feature ensures smooth gear changes, most pleasingly so in the upper echelons of that mammoth nine grand rev range. If you don't wish to use the rev-matching feature, just stay in Normal mode, or venture into the PCM to turn it off in Sport.

Another cool feature of both GT3s is the ability to flat shift, meaning you can keep on the gas while depressing the clutch pedal and reaching for a higher gear, only this time without destroying your clutch. This is because engine speed is automatically controlled by the car's ECU during this process, allowing for quicker gear changes, even if it's only by a few precious milliseconds. This feature may conjure some 'nanny state GT3' jibes, but in a car that revs so freely and so fast, my view is that any tech which can help us mere mortals keep up with it is most welcome. The reality is if you don't wish to use the flat shift feature then, well… don't!

On these roads, we spend our time switching mainly between second and third gears, only dropping to first for a super-tight right-hander, eeking out the revs each time to a cacophony of noise before changeups. On these roads, the cars are downright playful,

and it's a joy to drive them. A PDK equivalent would offer up a different kind of fun, but right now both Phil and I are pleased to be part of the 40 per cent of modern GT3 drivers who opted for a third pedal.

So, on paper, yes, the transmissions in these cars are the same. However, right where it matters – in the driver's seat – there are clear differences between the manual 991 and 992 GT3. I myself prefer the weightiness of the *shifter* in the 991, yet find favour with the fluidity of the *shift* in the 992.

Of course, there's no right or wrong answer here. As ever with Porsche, it's a case of choosing the right setup for you. Both GT3s are mesmerising machines and exquisite examples of automotive art, representing the high watermark of what just might, in years to come, be viewed as the golden era of the sports car, their manual gearbox at the heart of what makes them so special. ➲

HEAD TO HEAD
991 VS 992 GT3 MANUAL GEARBOX RATIOS

991 GT3

DRIVE SYSTEM: Rear-wheel drive
GEARBOX: Six-speed, GT Sport transmission; Porsche Torque Vectoring (PTV) including mechanically rear differential lock with asymmetric locking rate (30 per cent traction, 37 per cent overrun)

Gear ratios
FIRST GEAR: 3.75
SECOND GEAR: 2.38
THIRD GEAR: 1.72
FOURTH GEAR: 1.34
FIFTH GEAR: 1.08
SIXTH GEAR: 0.88
REVERSE GEAR: 3.42
REAR AXLE RATIO: 3.96

991 vs 992 GT3: battle of the gearbox | 73

992 GT3

DRIVE SYSTEM: Rear-wheel drive
GEARBOX: Six-speed, GT Sport transmission; Porsche Torque Vectoring (PTV) including mechanically rear differential lock with asymmetric locking rate (30 per cent traction, 37 per cent overrun)

Gear ratios
FIRST GEAR: 3.75
SECOND GEAR: 2.38
THIRD GEAR: 1.72
FOURTH GEAR: 1.34
FIFTH GEAR: 1.08
SIXTH GEAR: 0.88
REVERSE GEAR: 3.42
REAR AXLE RATIO: 3.96

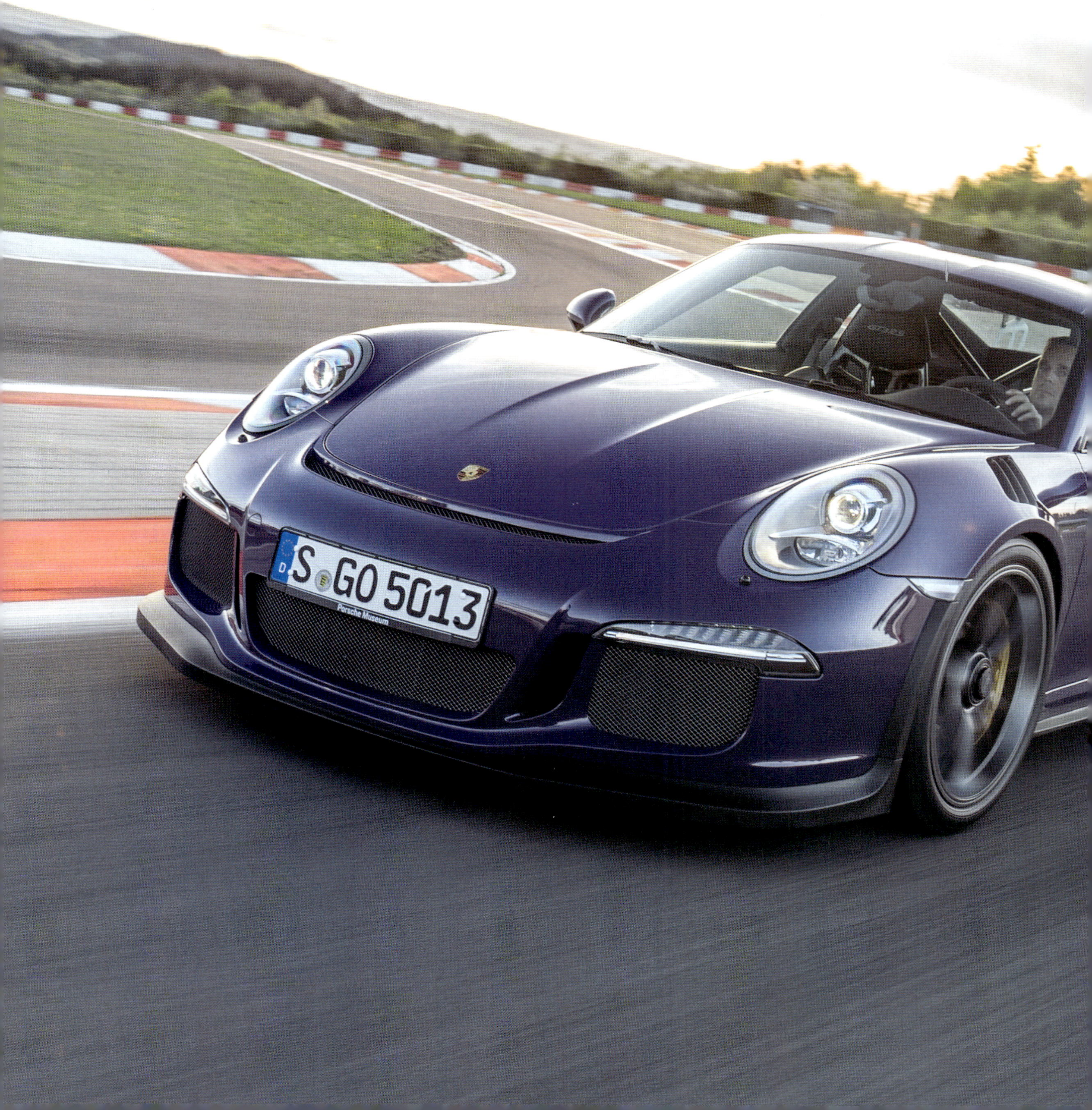
S GO 5013
Porsche Museum

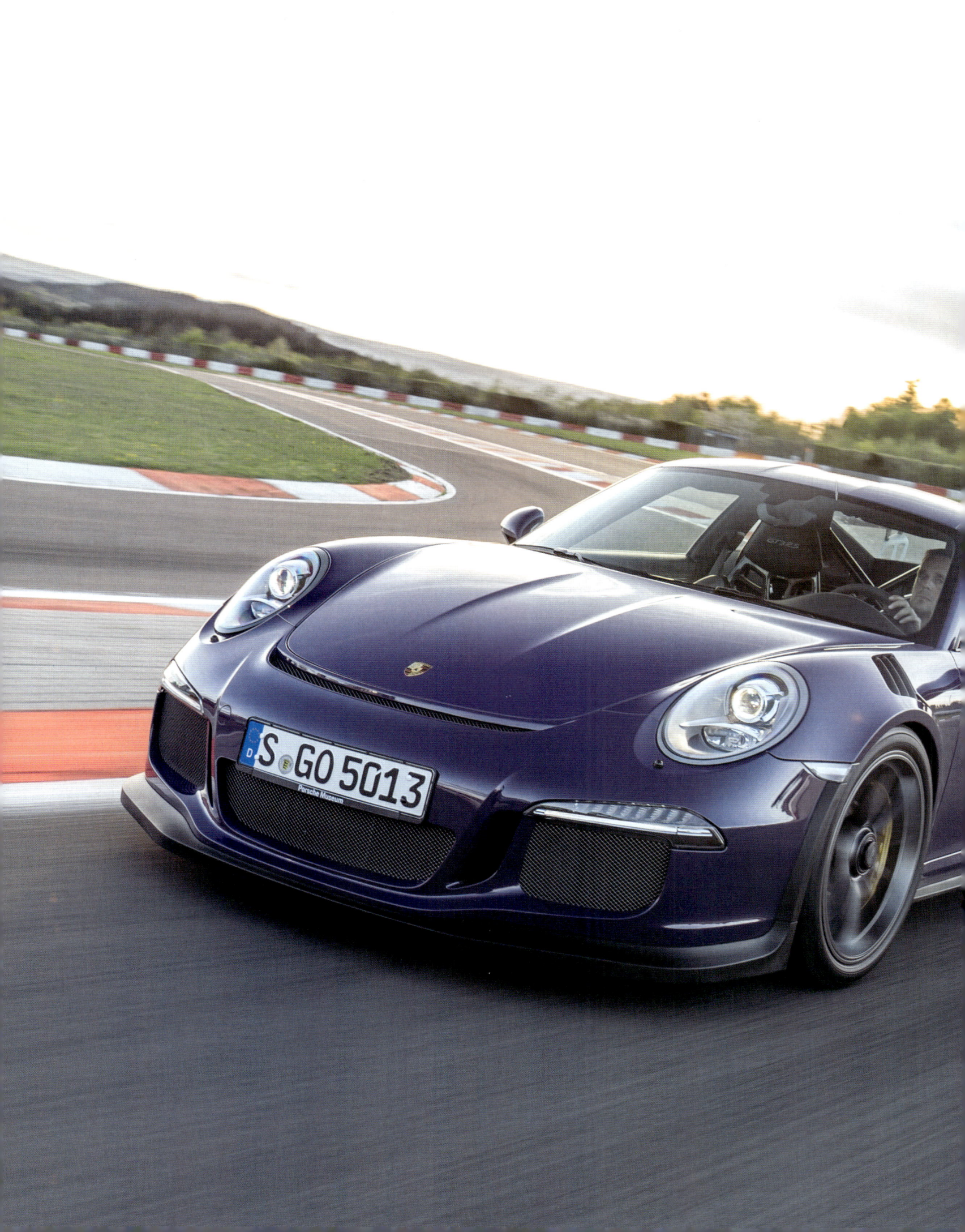

INDEX

76
996.1 GT3
Porsche 911 GT3 genesis: now a prized collector item, here's how to find a good one

84
997.2 GT3
The last Porsche GT3 with a fabled Mezger engine is a cracker to drive. Here's everything you need to know

92
997 GT3 RENNSPORTS
Want to add a 997 GT3 RS to your collection? We present everything you need to consider when searching for a great example

100
991.1 GT3 RS
It's the best-value RS you can buy, but there are plenty of pitfalls if you buy a bad example. Our guide ensures that won't happen!

Ultimate 911 GT3 Collection

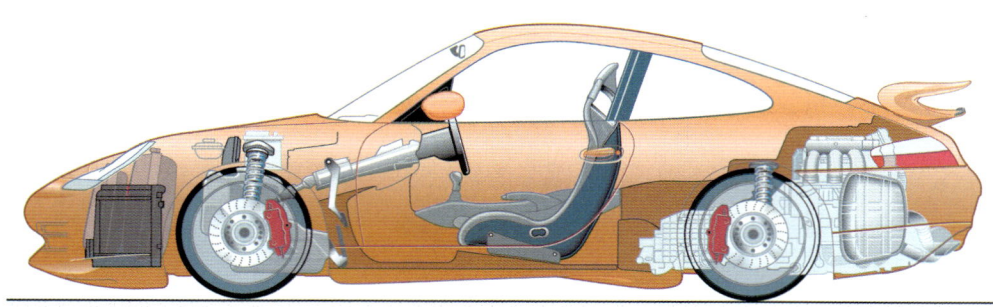

PORSCHE INDEX

Written by **Kieron Fennelly**

996 GT3

The 996 GT3 took Motorsport Porsche 911s in a completely new direction. More than 20 years on, both the Gen1 and Gen2 are highly sought after, so here's everything you need to know…

HISTORY & TECH

With the advent of the 996, Porsche decided to abandon GT2 racing and focus on GT3. The cars would be less expensive to build and expand the company's opportunities to race 911s in championship series in front of a much wider public.

Experiments with a prototype GT2 had also shown that even with 600hp, the turbocharged 911 would no longer be fast enough to beat GT2 competitors. Limitations of the M96 lubrication system (the engine fitted in the Carrera) meant that the new GT3 would require a bespoke unit.

The first GT3 was revealed in 1999 – a limited production run to qualify the race credentials of the 996. It differed from its air-cooled RS forebears: in addition to a specific engine, it used a heavier chassis, the more rigid shell of the Carrera 4 and was 40kg heavier than a stock 996 C2. Production realities meant that there was minimal scope in 1999 to make the GT3 lighter.

Despite its relatively impoverished state, Porsche didn't stint on an engine that's become known as the 'Mezger', even though Hans had retired in 1993 and had nothing to do with it. The credit is due to Herbert Ampferer, the long-serving mechanical engineer whose main claim to fame was the flat six that powered the Le Mans-winning GT1, a hybrid design with a fluid-cooled head.

Herbert returned to this idea, taking the aluminium bloc of the 964/993 and grafting a specific water-cooled 24-valve head on to it, with the capacity increased from the GT1's 3.2 to 3.6 litres. This was an expensive unit to build and even now the specification is impressive. The forged steel crankshaft had plasma-nitrided surfaces and titanium connecting rods that operated aluminium pistons in Nikasil bores. The head had a separate casing carrying the camshafts and tappets made from aluminium alloy, while the head itself used a special heat-resistant alloy with copper, zinc and trace elements of antimony, cobalt and zirconium.

Engine coolant travelled through an oil/water heat-exchanger on the crankcase. Four scavenge pumps, one in each cylinder head and two in the sump, circulated the oil through the engine and into a separate reservoir. A more accurately monitored Variocam mechanism, operated hydraulically rather than using the 996's chain-drive, responded to commands from the ECU by either advancing or retarding the intake camshaft. With the GT3's 355bhp (360PS) now at 7,200rpm, the road-going car was almost in the same state of tune as the R and Cup track versions.

Some 1,858 GT3s were made between 1999 and 2000, while a further batch of 2,800 – the Gen2 GT3 – appeared in 2003/4. These can be identified by the revised headlights and larger fixed rear wing. Not so apparent were the substantial changes made to the 3.6, which now met Euro 4 and for the first time, US emissions standards. Variocam Plus and a significant reworking of the head, including lighter connecting rods and valve gear-enhanced breathing, meant that power output was now 376bhp at 7,400 rpm. The rev limit was increased from 7,800 to 8,200rpm. ➲

THE VALUES STORY

It took some adjustment for Porsche fans to accept that the new GT3 was no lighter than the Carrera and ostensibly didn't accelerate any faster than the (suspiciously quick) 996 3.4-litre press cars. Yet as the first road tests revealed, this lower, harsher-riding 911 with "race car technology and stunning engineering" as *Autocar* put it, was at £76,500 an amazing performance package when the Carrera cost £65,000. The entire production sold out within weeks and when the second-generation GT3 appeared in 2003, it too flew off the shelves.

Rarity and demand kept depreciation away and it was only after the launch of the 997 GT3 that values fell significantly below £70,000 to a low around 2015. But if today the 996 GT3 is an old car and subsequent GT3 generations are much more sophisticated and – on the track – far quicker, nevertheless as a rare 911 the original GT3 is still worth more than its 1999 retail price. At reputable dealers the current level, which has been unchanged for some time, sits between £74,000 and £95,000. Offers much below this bracket usually involve examples that either have a significant accident story or very high mileage, although a 120,000+ miles car with a consistent history shouldn't put off a knowledgeable buyer.

ABOVE The GT3 had broadly the same 996 interior, albeit with a higher redline, and without a lower centre console

MARKET RIVALS

The £70,000-£100,000 bracket opens up a field of enticing alternative 911s

964
In some ways just as uncompromising as the 996 GT3, the last of the frog-eye 911s trades a less-than-boulevard ride for a visceral, vintage Porsche driving experience and a splendid exhaust note.

991.1 GTS
The same money buys a 40,000-mile, 2015 example of the last nat-asp 911. GTS-spec brought the Powerkit, upgraded cabin and other desirable extras. Massively fast and eminently trackable, it's not a racer in the GT3 sense, but a less-demanding drive when you want it to be.

993 C2S
Long considered the most desirable of the 993 range, the 2S carries a significant premium over the C2 and the best sell well into six figures. A five-owner manual example is on offer from a private seller in Woodford, east London, at £95,000, but it does require minor recommissioning.

992 C2S
The current 911 S offers the only a manual gearbox 911 option (excluding the 4.0-litre cars). Two-year-old examples abound between £90,000 and £100,000, but depreciation will be steep. An easy 911 to drive, especially with PDK, but many enthusiasts may find it rather bland.

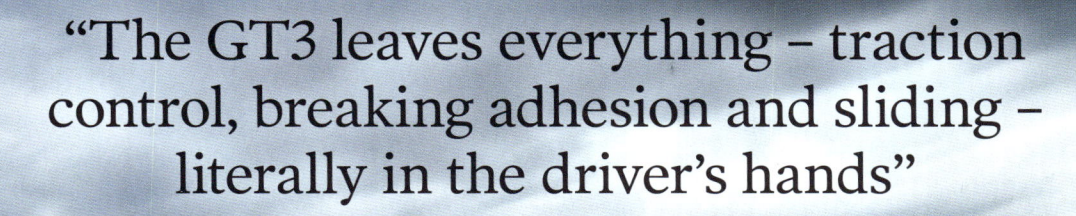

> "The GT3 leaves everything – traction control, breaking adhesion and sliding – literally in the driver's hands"

ABOVE The 996 GT3 was designed to homologate the 911 for endurance racing, and the public got its first taste of the car when it was revealed at 1999's Geneva Motor Show

DRIVING EXPERIENCE

The GT3 was never for the faint hearted: the driving position alone with specific sports bucket seats and cabin denuded of rear fittings already suggested a driver-oriented machine. Outside, a double-wing spoiler and deeper front splitter added to this impression and sitting 30mm lower than the C2, the GT3 clearly meant business.

Once underway, the looks of this Weissach-inspired 911 are not deceptive: the driver has no alternative but to take charge. The flat six is rarely temperamental, but it longs to rev, urging the car on. Driving the GT3 requires a conscious approach. Clutch, gearshift and steering are all heavier than a Carrera's, but deliberately so: the precision of these superbly analogue controls is staggering, the steering especially. With only three turns between locks it's quick, and in combination with that extraordinarily reactive flat six the GT3 was a track car adjusted for the road, not vice versa.

On an open road with twists, turns and gradients, the GT3 is in its element. If not quite as lithe (inevitably) as its fabled ancestor the 2.7 RS, it corners with all the accuracy the committed driver would hope for, with almost no roll and without the understeer that builds up as the going gets quick in the Carrera. Meanwhile, the flat six responds to the slightest depression of the throttle pedal, the aural accompaniment alone encouraging the pilot.

While this is classic 'man and machine' symbiosis, it requires a sensitive and experienced man or woman at the wheel. Apart from effective ABS and a locking differential, the GT3 leaves everything – traction control, breaking adhesion and sliding – literally in the driver's hands.

ABOVE Leather-clad bucket seats were standard equipment for both the Gen1 and Gen2. A Clubsport package had these trimmed in fire-retardant material, while a rollcage and extinguisher was also fitted

BUYING A 996 GT3

History is never more vital than with the GT3. Potential purchasers need not be put off by a car with significant circuit mileage, but in that case, it needs evidence of the more intensive maintenance regime that Porsche recommended. This included more frequent engine oil changes, brake fluid and coolant replacement, as well as fresh oil in the gearbox at the end of a track season.

All these engines were built for racing and a car that's been tracked yet properly looked after will often run better than a very low mileage GT3 which, because of infrequent use, may not always have been serviced annually. Note that the track car will show more wear. Oil weeps, rarely significant, can occur around the timing chain housing. Serious leaks are unusual and would demand further investigation.

The state of the coolant system is critical: is the pipework original or has it been replaced? Are the coolant radiators corroded? Perspective matters, though: Ray Northway, who's worked on the "rattly old Mezger" as it's affectionately referred to, says that considering its 100bhp+/litre output, it's extraordinarily rugged. Nevertheless, consumables do wear out: suspension components, bushes, dampers and even arms don't last 20 years. Neither do coil packs, which expire especially when exposed repeatedly to road filth. Brake pipes replaced with braided hoses are a worthy improvement and a sign of thorough maintenance. Clutches were reckoned to last about 70,000 miles and LSDs on a track-rich diet can become noisy at half that mileage.

Porsche bodywork doesn't normally corrode unless accident repairs have been substandard: the GT3 buyer will be looking for signs of severe accident damage. Many GT3s will have had their front and/or rear bumpers replaced, but this goes for a lot of 911s, and if correctly fitted, it shouldn't affect the car's value.

Mark Sumpter of Paragon has sold GT3s since the early days. He always pulls up the carpet in the footwells to see evidence of welding, explaining that ripples in the floor are a sign that the chassis has been straightened on a jig. This is an indication of a more serious 'off'. For many would-be buyers, an expert assessment – particularly of the underside – may be a wise investment if purchasing outside the network of specialist Porsche independents.

Some owners modified their GT3s with (usually stiffer) non-standard suspension parts. This spoils ride on the road and rarely improves on the original equipment. The cost of restoring the factory setup should be factored into the discussion. As with all 911s, highly customised GT3s are to be avoided unless the buyer is planning a total rebuild and the price reflects this.

DESIRABLE OPTIONS

The 996 GT3 is a purpose-built, driver-focused package with a stiffer, non-adjustable suspension, specific exhaust and its own gearbox that was derived from the 450bhp 993 GT2. The options list was quite short – a few cars will have been specified with their sound systems and/or air-conditioning deleted.

A more important consideration is whether to seek out a GT3 with the no-cost Clubsport cabin. The CS versions had a single-plate, lightened flywheel and its lack of inertia can make it a chore in stop-start traffic. The fitted half-roll cage stiffened the shell and imbued confidence on the track, but within the cabin it can be rather intrusive, reducing luggage access to the back that's already awkward when the non-folding bucket seats are fitted. These seats incidentally can feel rather tight for larger frames.

On steel discs, the GT3's stopping power was judged "phenomenal" by *Autocar*; carbon ceramic brakes were a controversially expensive £5,800 optional extra for Gen2 cars. Stumping up a further £3,875, a few buyers specified the full leather/carbon-fibre package for the cabin.

INVESTMENT POTENTIAL

The 996 GT3 was both the first of its type and a low-volume 911. As such it's never really lost value. Purchased today in the bracket suggested, an authentic, historied example should continue to appreciate quietly while remaining eminently usable. Indeed, these thoroughbreds need regular exercise to stay fit.

As an investment, the 996 GT3 faces competition from the 997 variety that had a better ride and was a more resolved 911, but the pioneering aspect of the earlier car, plus the fact that the Gen1 was the last road-going Porsche built on the Weissach production line, simply underline its historical significance and value. Some advertisers make the fanciful claim that the Gen 1 car has special value because it was 'hand built'. Yet the reality is that the second edition was an improved offering with not just a more developed engine, but because it had come down the normal production line, benefitted from experience with the earlier version.

On the road, Gen 1 and 2 differences are slight. With 30 or 40hp more and a higher rev limit, the later, more stiffly sprung GT3 might feel a shade quicker, but their worth today depends more on condition and history rather than type. **911**

BELOW The Taco-winged 996.1 is the only GT3 with no RS model ahead of it

TOTAL 911 VERDICT

Especially with a GT3, the buyer needs to be certain of what they're acquiring. As one of the most uncompromising of 911s, many GT3s have had multiple owners – often because buyers underestimated this demanding 911 and traded it again. Yet the harsh, low-speed ride and heavy-seeming controls all have their purpose: from the off, the GT3 responds best to a driver who's able to empathise with it and give it its head on every journey, which alas on much of this crowded and semi-urban island is simply not feasible. Unless you live in remoter parts, the very analogue 996 GT3 is not a regular road driver, but given a suitable environment and in the right hands it's undoubtedly one of the most satisfying and engaging sports cars ever built. Unless used regularly on the track (and few appear on circuits now because 20 years on they're deemed too slow by many track devotees), maintenance costs are no more than a Carrera's. These days attracting the collector sphere, for the driver with access to decent roads and who relishes a challenge, the original GT3 makes a fine driver's car… perhaps more for a 50-mile blast than a 500-mile trip.

★★★★☆

PORSCHE INDEX
997.2 GT3

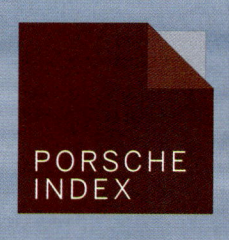

Written by **Kieron Fennelly**

A minor engineering masterpiece, the 997.2 GT3 makes for a dynamic weekend car that needs neither heated garage nor mollycoddling, and takes to the track with aplomb

HISTORY & TECH

With the advent of the 996, Porsche decided to make the new GT3 category its competition arena. Previously, it had campaigned in GT2 and although the 993 GT2 was moderately successful, as mid-engine opposition became stronger Porsche knew it was at an increasing disadvantage with the rear engine.

The switch to GT3 was entirely logical. As a simpler formula, the Porsche GT3 was less expensive to build and could be sold to far more club racers. More racers meant greater exposure to the public and a virtuous circle was thus created. It was an inspired move and almost three decades later has spawned many more race series and competitors from other manufacturers.

Whereas in air-cooled days the RS had served as the homologation basis, Porsche designed the 996 GT3 as its racing model. It came with its own engine, popularly honoured with the name 'Mezger' (even though Hans Mezger had long since retired) because the flat six was a hybrid of the air-cooled 964-993 block with a bespoke water-cooled cylinder head. Restrictions meant that the competition 996 R was only marginally more powerful than the 360PS road-going GT3, and stiffly suspended. Both generations of 996 GT3 were uncompromising as road cars.

The 997.1 GT3 that was launched in 2006 was more powerful. Yet with adjustable suspension and electronic control of the chassis – something that was lacking in its predecessor – it was also a more comfortable and sophisticated road car.

The major change with the Gen2 model three years later was the larger engine. Bored out to 102.7mm (generating 3,797cc) the now 3.8 effectively homologated the engine, which the competition GT3s had been using since 2005.

A noted and much admired characteristic of the 'Mezger' engine was its amazing high-revving ability, enhanced further with the 3.8 edition. Not only was it marginally more powerful – 435PS, up 20PS over the 3.6 – but it also generated more torque, which increased from 405 to 430Nm. These striking figures were achieved at yet higher rpm, with maximum torque reached at 6,250rpm and maximum power at 7,600rpm. Interestingly, Porsche said the engine was safe for 8,500rpm.

Whereas non-PDK 997.2 911s used a six-speed 'box from Aisin, transmission for the GT3s remained in the hands of the robust Getrag six that was originally developed for the 993 GT2. The Gen2 GT3 also featured suspension and chassis upgrades. Once again adjustable, Porsche's Active Engine Mounts joined the two-stage PASM and a lowered front roll centre using the softer anti-roll bar of the GT2, combined with the GT2's stiffer springs. This all contributed to sharper handling. PSM could be turned off if the driver wished. A new, larger wing that had been developed from racing brought greater high-speed stability, doubling downforce at maximum speed.

With these enhanced specifications, Walter Röhrl was able to lap the Nürburgring in 7:40 against the 7:45 he'd managed in the 997.1. Time however, marches on: in 2023 the 4.0-litre 520PS 992 GT3 lapped in 6:55.

Ultimate 911 GT3 Collection

THE VALUES STORY

The Gen2 997 GT3 appeared in 2009, priced in the UK at £82,000. Once again a 911 GT3 offered remarkable value for money when a well-specified Carrera S cost almost the same. Demand for the 3.8 GT3 was significant, factory output limited and a few buyers paid over the odds rather than join the waiting list. The model hardly depreciated until the arrival of the 991 GT3, which was delayed until late 2014 because of engine problems and by then all Porsche prices were on the rise.

The high point for the 997.2 GT3 was probably reached in 2018. Jamie Tyler of Paragon Porsche recalls selling a black 997.2 GT3 with 11,000 miles for £115,000. He estimates that today he might pitch that same car showing, say 40,000 miles, at between £90,000 and £100,000. The few 997.2 GT3s advertised in spring 2024 suggest that the going rate is between £85,000 and £100,000. A rare model, with perhaps fewer than 200 in RHD, means owners seem to be holding on to them. Tyler thinks Paragon hasn't sold one for some years, whereas the RS version does pass through its hands from time to time.

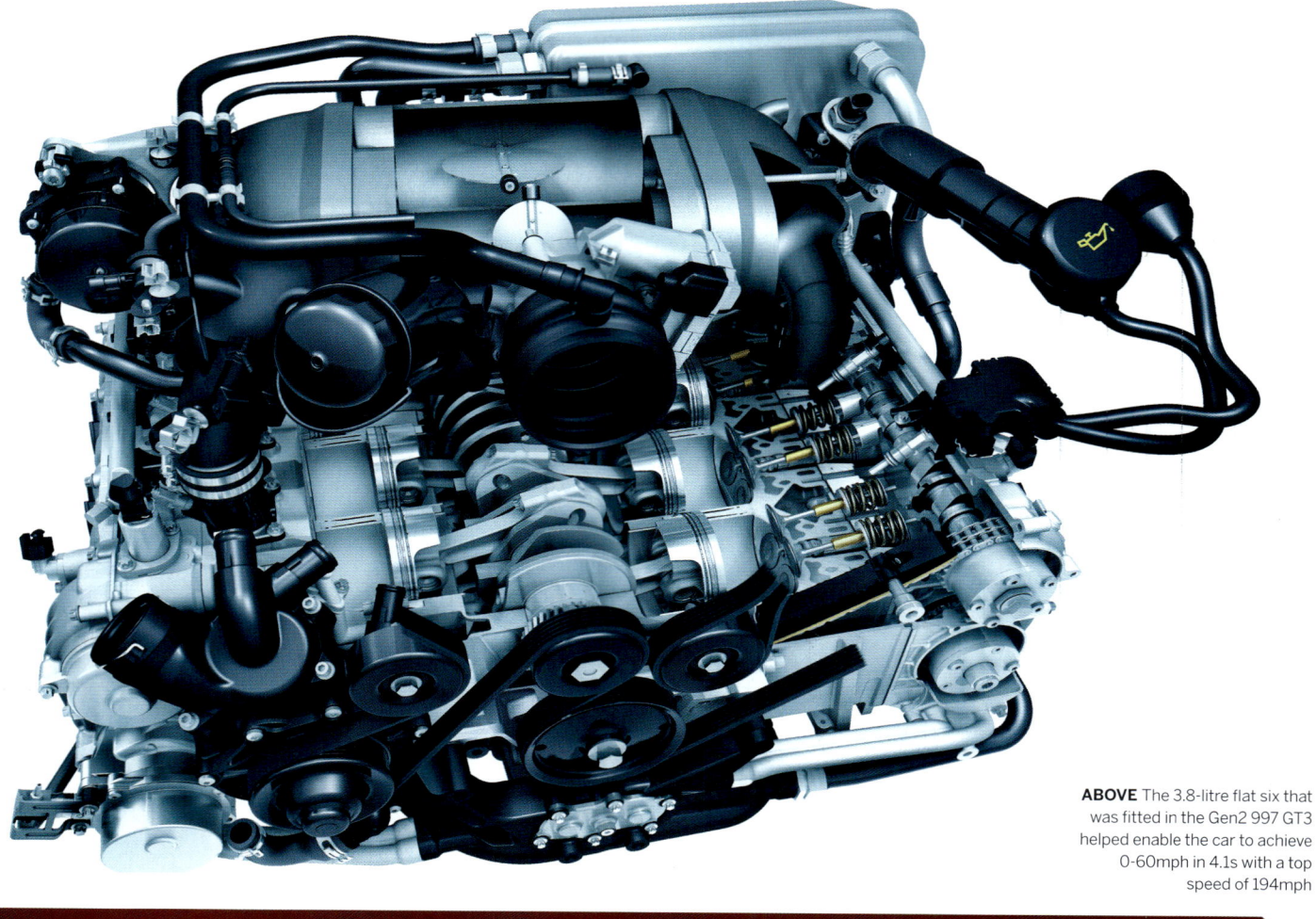

ABOVE The 3.8-litre flat six that was fitted in the Gen2 997 GT3 helped enable the car to achieve 0-60mph in 4.1s with a top speed of 194mph

MARKET RIVALS

A budget of between £85,000 and £100,000 throws up some interesting 911 possibilities...

997.1 GT3
The rarer first-generation 997 GT3 has a slightly less-developed chassis but is just as quick. Ray Northway is offering a FPSH Comfort spec 2006 example with 49,000 miles at £83,995.

991.1 GT3
With PDK and more sophisticated electronics, the 991 is dynamically on another level. It may be larger, but it's less demanding to drive thanks to a much higher level of tech. No shortage on the market, with most priced around £95,000.

991.1 Turbo S
With its colossal performance, the PDK-only Turbo S is probably the most accessible and best-built grand tourer there is, yet it's also quite capable on the track. There's no shortage of examples around £100,000, but they're still depreciating significantly.

964 Carrera
The 964's ranks were plundered by the restomod industry, but is now a sought-after 911. Raw, with all the traditional 911 virtues and yet with modern underpinnings, the 964 has more usable performance for UK touring or Sunday runs to Goodwood.

"Few production cars have ever been so uncompromisingly driver-oriented"

DRIVING EXPERIENCE

The press launch, involving a select band of Porsche scribes, took place in the hills south of Stuttgart in May 2009. Racer and author Tony Dron sought out the nearest Autobahn and proceeded to praise the new GT3's "perfect" stability at 180mph. Jamie Corstophine, then writing for *Autocar*, went into raptures about the GT3's steering: "Beautifully weighted, perfectly geared and alive with information." He felt that Porsche's changes to the front suspension had worked, finding better turn-in and less understeer. Veteran commentator Andrew Frankel observed that, "What you notice most is not its raw grip, but the willingness of the GT3 to bend itself to your will. It not only goes where you point it, understeering or oversteering like any proper Porsche, but the scope it provides you to control these forces is wider than any Porsche I have known." Other correspondents remarked on the ride (on Germany's admittedly smooth blacktop), saying that it was notably softer and easier going than earlier GT3s.

Today's driving experience is unchanged. The extraordinary hydraulically assisted steering, the instant throttle response… few production cars have ever been so uncompromisingly driver-oriented. At first, the controls feel heavy, largely because the tendency for all manufacturers has long been an unnecessary over-servoing of both brakes and steering. With the GT3 (as with any 911 but more so) it doesn't take long for a driver with any sensitivity to realise that what at first seems heavy is the correct weighting.

The GT3 won't drive itself, though. It requires constant driver input that takes practice. Unlike the 911 GTS, say, where after a track session the driver can relax, the GT3 demands concentration and would suit few people as a commuting Porsche, especially in Britain where its 'Autobahn' gearing is too high, although this criticism applies to other manual Porsche, too. However, on the right road, the combination of response, absolute precision, visibility and sensational soundtrack makes the GT3 an almost indescribably satisfying experience.

BUYER'S GUIDE

The GT3 is a specialist car and the best source is specialist dealers. Often, these Porsche independents know the car and have sold and serviced it before. They're also discriminating and decline to handle certain examples. Paragon, for instance, won't buy cars with service gaps of three years or more; Ray Northway is always careful to avoid a Porsche he thinks is at all dubious. Expect to pay less for a purchase from a private seller, but greater caution is required. Ideally, the seller would be a Porsche Club member. The GT3 community is quite small, and owners and cars are often known to each other. A GT3 offered at a very low price (rare, but it is possible) should elicit immediate suspicion.

The phrase 'bullet-proof' often occurs when discussing the 'Mezger' with the specialists. What these engines lack in refinement, they make up for in ruggedness. Mechanically speaking, little goes wrong. The valve chains can stretch and rattle, noticeable below 3,000rpm, hence the engine's 'rattly' reputation, but Northway's Paul Stacey says this isn't a reason to replace them. A GT3 is hard on its drivetrain: the clutch can start to feel heavy after 30,000 miles and synchromesh on 2nd and 3rd on gearboxes that have seen significant track use can weaken. Porsche's limited slip differential will betray signs of fatigue by groaning during parking manoeuvres.

On what are now old cars, other deterioration is often age- and wear-related: cooling systems, exhaust flanges and fittings, suspension arms and other chassis components. A correctly serviced GT3 will have had much of this replaced at some point. These cars tend not to cover huge winter mileages, so the underside corrosion that afflicts other well-used 911s should be less apparent. The same goes for the body, and any signs of rust are attributable to inadequate repairs. Few GT3s of this vintage will not have had at least a respray of the front bumper, if not its replacement by now. Like all 997 cabins, the interior wears well. However, Alcantara surfaces can look rather flat after a time and the outer bolster on the driver's seat can show wear. This is usually the sign of either a higher mileage or multi-owner car. ➲

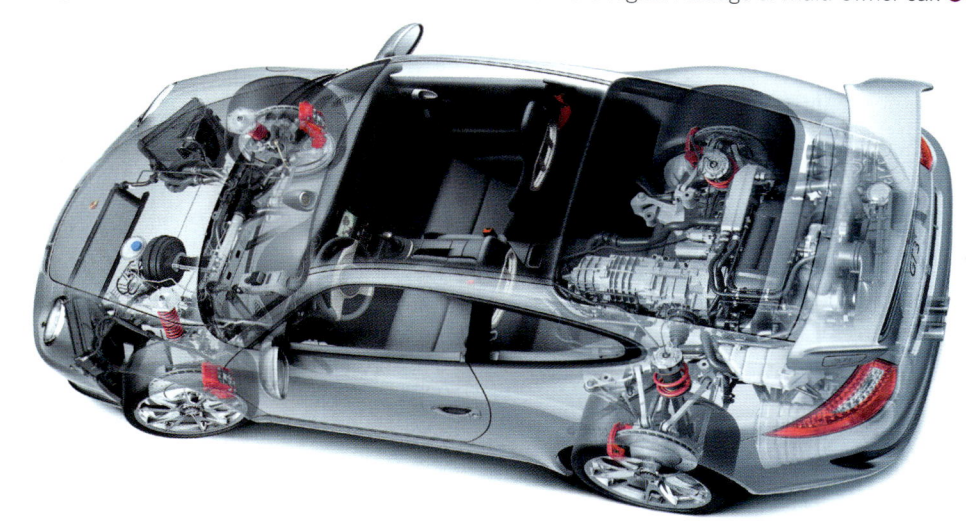

DESIRABLE OPTIONS

It's significant that today GT3s with the Comfort cabin are, according to Paragon, the more sought-after model, which indicates the kind of use owners now expect. The Clubsport's half roll-cage and six-point driver's harness are cumbersome for casual use, while the lightweight bucket seats can be a tight fit for larger individuals. As usual, Porsche's carbon ceramic brakes were the most expensive option, saving 10kg per wheel, but again given a road-biased diet, PCCB is of less interest today. Some earlier cars weren't fitted with the practical front axle lift – this is worth looking for – and one or two (misguided) original buyers ticked the 'A/C delete' option.

INVESTMENT POTENTIAL

In contrast to later GT3s, the 997 cars were produced in small numbers with possibly as few as 200 997.2 GT3 in RHD. Their market value now is similar to or slightly above their original selling price. Unlikely to be purchased today by serious track devotees, the 997.2 GT3 not only appeals to collectors, but also to 911 enthusiasts happy to use this thoroughbred for high days and holidays, while taking it to the track on occasion. In contrast to the early, air-cooled cars, absolute originality matters far less for the modern 911s – many GT3s will sport replaced splitters or front bumpers. High-mileage examples as ever are worth slightly less, but the distinctly analogue and 'last of the line' nature of the 997 GT3 means it should continue to appreciate modestly.

"Absolute originality matters far less for the modern 911s – many GT3s will sport replaced splitters or front bumpers"

TOTAL 911 VERDICT

The 997.2 GT3 represents the end of an epoch, and brings with it a raft of notable 'lasts'. In Porsche terms it was the last 'Mezger' engine 911 (excluding, of course, the collector-only 4.0 RS). This was an unrefined yet amazingly robust unit that despite yielding well over 100bhp/litre, was capable of immense mileages – 200,000 miles even before rebuilds became necessary. It was the last 911 that didn't need parking sensors or other aids to visibility, and relatively basic chassis electronics aside, the last analogue 911. It was also the last classic GT3 – characteristics that today appeal to the collector who's also attracted by its rarity, or the real enthusiast. However, if you fall into the latter group, you need to ensure that this 911 is for you.

The 997 GT3 is an intense 911. It's either on or off: there's no 'cruise mode' and the driver will need at least 75 per cent concentration in most situations. Sound levels are high in the cabin. They'll drown out the radio, but on a twisting road, on anything other than a cold day, the only accessory a driver will need is the air-conditioning! Possibly the GTS – a little-used manual 997 GTS (still with 408PS) – is a less-demanding partner, and for around 20 per cent less, might be a safer bet. However, if the 997.2 GT3 does fit your driving tastes and lifestyle, then virtually depreciation-free, it's a purchase you're unlikely to regret.

★★★★☆

997 GT3 RS

Written by Kieron Fennelly

PORSCHE INDEX

Representing the final analogue era of RS, for enthusiasts the 997 has already achieved classic status. Total 911 documents the Gen1 3.6 and Gen2 3.8

HISTORY & TECH

On the water-cooled cars, where the GT3 had taken over the RS role, the RS was reintroduced as the homologation version. Porsche's first foray was the 996 GT3 RS that revived the tradition of lighter body panels, harder suspension and a more powerful engine. It also used the wider Carrera 4 body. This, in Andreas Preuninger's words, was a parts-bin exercise "to see what we could do." Today, Andreas – the director of GT Cars at Weissach – admits that the 996 GT3 RS needed more development, but the car covered its costs and the exercise provided the template for the more sophisticated 997 GT3 RS, the first generation of which would appear in spring 2007.

The 996 GT3 used Herbert Ampferer's 3.6, a development of the Le Mans GT1 engine that allied the 964/993 crankcase with a bespoke, water-cooled 24-valve head. This was officially rated at 376bhp for the Gen2 996 GT3, although in reality it was nearer 400bhp for the RS.

Incremental improvements for the 997.1 GT3 – lower reciprocating mass, compression raised to 12.1:1 and a new induction system – contributed to a power output of 411bhp at 8,400rpm, almost 115bhp/litre. Porsche stated the engine was safe for 9,500rpm, which enabled some margin for premature downshifts.

The same specification was employed for the 997 GT3 RS. Like the 996 RS, the 997 edition used the Carrera 4 body. This made it wider than its GT3 sibling, an impression reinforced by its larger, high rear spoiler and low front splitter. Suspension changes over the GT3 included rear suspension that could be adjusted for camber and lighter, forged uprights at the front. However, like the GT3, the RS version had PSM and PASM. A plexiglass rear window and carbon fibre boot lid brought minor weight savings, which enabled Porsche to claim a 20kg saving over the GT3 if the air-conditioning was deleted.

The second-generation 997 GT3 that appeared in 2010 gave Weissach the confidence to build the homologation version of the 3.8-litre race engine it had been campaigning with the 996 RSR. With a precise capacity of 3,797cc, for the Gen2 RS its compression ratio was higher than the 3.6's and power output was now 430bhp. Recalling the adverse comments on the Gen1 RS, this time Porsche ensured the RS would offer more for the £10,000-plus difference between it and the GT3. Not only did the second-generation RS feature an altogether more substantial aerodynamics package than the Gen1, it also sported a slightly symbolic 20bhp power uplift.

In 2011, the 3.8-litre RS gave way to the 4.0-litre RS, a last hurrah for Ampferer's hybrid design, now rated at 493hp, which everyone still insists on calling the 'Mezger' engine. It featured, among other things, titanium con rods and a lightweight crank taken from the RSR race car. ➲

THE VALUES STORY

The 997 GT3 RS followed the launch of the GT3 in 2007. In the UK it was priced at £94,280, some £14,000 more than the GT3. The significant price difference caused some surprise because performance claims were the same. At its launch the 997.2 GT3 RS was presented in 2010 with a £104,841 UK retail price. Their rarity – 1,106 RS from the first generation and 1,619 Gen2s – combined with their distinctly specialist nature, meant these cars never depreciated in the same way as production 997s or even the GT3. The lowest point was probably in 2012, when the Gen1 was selling at around £70,000. Today, prices begin at about £130,000 for a well-used example, to over £200,000 for the most expensive Gen2. Usually, the earlier car occupies the sub-£150,000 sector, but pricing also depends on condition and usage.

The much-heralded 4.0 RS appeared in 2011, and virtually all 600 cars had sold before they left the factory. Collectors' cars from the outset, RHD 4.0 RSs were resold within months at well over twice their £128,466 retail cost. Since then, the 4.0 RS has remained firmly in the collector sphere. In 2021, a RHD example changed hands in Britain for £430,000; on the continent, LHD prices are still well above €300,000.

MARKET RIVALS

The GT3 RS has almost no rivals among other Porsches. The owner has to want an RS for what it can do over and above other 911s, or they're a well-heeled collector, in which case interest lies with the 4.0 RS.

992 GT3 RS
Recently launched, the 520bhp RS must, given the black clouds of environmental correctness, be the ultimate expression of the internal combustion 911. It's listed at £180,000, but is virtually unobtainable, so is likely to be offered on the resale market for more than double.

964 RS
Twenty years ago the track day favourite, but today considered too valuable, the 964 RS offers visceral, usable performance for more average drivers not in a position to exploit a 997 RS fully. The 964 RS sells today from £150,000 to over £200,000.

2.4S
Another classic 911 now going for over £150,000. Not quick by today's standards, the early 911 offers a wonderfully involving vintage driving experience (and below the speed limit) that's unmatched by more modern, complex and heavier 911s.

997 GT3
Do you really need the RS aero? The narrower GT3 is almost as good on the track and much more practical for daily use. An exceptional 997.2 sells for around £120,000, while a moderate mileage 997.1 can be found for around £85,000.

DRIVING EXPERIENCE

Porsche sought no excuses for making its 997 GT3 RS a challenging drive. Not because of any dynamic shortcoming, but because the RS was a finely honed precision tool that demanded measured and considered inputs from its driver. Today, as then, control weights initially feel heavy: the clutch is firm (PDK wasn't offered until the 991 GT3 in 2013) and the short shift gearlever requires a conscious flexing of the wrist. It requires a particular style, but the keen driver quickly adapts, and directing the RS becomes both second nature and utterly absorbing.

The hydraulically assisted steering, which at the outset feels both heavy and over-sensitive, reveals itself a faithful servant entirely responsive to the driver while continuously transmitting tactile information on the road in a way that with electrical steering we've largely forgotten. The throttle too, requires careful application: its sensitivity is much increased by the single mass flywheel that chatters and rattles at idle.

Little needs to be said about the performance of an RS. One step removed from a pure race car, what's amazing about this Porsche is not its 9.5 seconds to 100mph or a top speed of almost twice that, but how such a guided missile has been made tractable enough to drive in urban traffic. In a review for *Autocar* in 2007, Chris Harris decided that only the 997 RS's ample girth – too wide for some of Britain's narrower B roads to be tackled comfortably – would discourage him from everyday use of this 911. He certainly felt that the adjustments to the suspension made the RS a palpably more secure car than the softer-sprung GT3, but its sheer energy was too much for ordinary roads and an owner would really need regular track dates to enjoy anything of its dynamic potential. ➔

BUYING A 997 GT3 RS

Even by Porsche standards, an RS is a specialist purchase. Many owners will be diehard enthusiasts and all cars on the market should have full histories. Despite its 110+bhp per litre state of tune, both 3.6 and 3.8 engines are robust, but regular servicing is vital and should be on a time–as–much–as–mileage basis. Cars that do 2,000 miles annually should receive almost as much workshop time as cars that cover 10 times that, and some do: in 2022 Luc Lecudonnec told *Flat 6* he bought his 997 GT3 3.6 new in 2007 and has toured 20,000 miles every year, driving to track days at Portimao, Nürburgring and Silverstone. Apart from breaking a pinion at 100,000 miles his GT3 has never failed and at 266,000 miles, he still sees no reason to change it. For peace of mind, he had the engine rebuilt at 260,000 miles.

Luc's experience underlines the crucial role of proactive maintenance: changing parts before they're life-expired and driving with technical sympathy – warming the engine carefully (not exceeding 3,000rpm) until the oil is at least at 60°C and always cruising the last laps of a track session to cool brakes, suspension and engine gently. This is the sort of treatment the would-be buyer should look for in any RS or GT3 service history, but these cars are rugged and some had service intervals skipped. Look for examples that have received preventive maintenance.

Even more than the GT3, the RS was made for the track and inevitably some cars will have hit barriers and needed replacement bumpers. Ensure that the aerodynamic parts that play an important downforce role in ensuring stability have been installed correctly. The suspension has a hard life heating up and cooling down, and a four-wheel geometry check is essential. Paragon Porsche knows these cars well and will take up the carpets and examine the floor of any GT3 or RS it suspects of having suffered a serious impact. It may even reject a car with signs of creases or chassis welding. ○

"The RS was made for the track and inevitably some cars will have hit barriers and needed replacement bumpers"

DESIRABLE OPTIONS

With its Alcantara seat fittings and steering wheel, and the rear compartment filled with a roll cage, there wasn't much to add to the 997 GT3 RS except perhaps more fire extinguishers. The Gen2 had the lifting front axle option and because the front spoiler is very low, this is a worthwhile addition. Air-conditioning was optional because its absence meant Porsche could claim a minor weight-saving over the plain GT3, and cars with air-conditioning are also worth seeking. The obsession with weight was also a reason for not fitting a radio.
Colour choice plays a part on value and desirability: the original Viper green is less ostentatious than orange with the black RS swoosh, and some buyers will prefer it. However, a Paint To Sample car will command a considerable premium.

| Ultimate 911 GT3 Collection

INVESTMENT POTENTIAL

The low-volume 911s have been a fair investment for some years. Some types appear to have levelled off, but the best 997 3.8 RSs have now passed the £200,000 mark and given the way the market is still buying into Porsche, pristine and little-used cars have the best potential. Average examples with good histories are also likely to continue appreciating quietly.

At Paragon, Jason Shepherd's view is that interest in the older models (997 and earlier 911s) is increasing because driving generally is becoming more automated and the keener driver is seeking out simpler, more analogue cars. The 997 GT3 RS's successor, the 991, represents too great a technical jump for many 911 enthusiasts.

The 4.0-litre 997 GT3 RS is in another category: the vast majority seem destined to continue reposing in their heated vaults, their owners hopefully awaiting the day when some buyer breaks the half-million Euro barrier... **911**

TOTAL 911 VERDICT

Engineering director Horst Marchart said they made the (original) GT3 the best they could and priced it as low as they could. It was a philosophy that's endured: compare the price of a new RS 3.8 in 2010 with the 997 Sport Classic. That the latter cost £30,000 more is instructive, suggesting a rather thicker profit margin than the RS. As a result the GT3 and RS have always been seen as especially good value, and bought for track use by some punters whose budget didn't extend much beyond the initial purchase price. Consequently, there are RSs that have been under-serviced, and damaged and badly repaired. These are the cars to avoid and will be at the 'bargain' end of the price spectrum.

Yet even more than a GT3, a 997 RS isn't for the faint hearted. More compact than the 991 RS, its electronics are less sophisticated and the car is potentially more challenging on the track than the later 911; some feel it requires more expertise to reach the limit. However, given that 997 RSs are now an average of 15 years old, this is now perhaps a rather esoteric consideration for would-be buyers, who are more interested in an analogue, earlier GT3 RS and driving fast rather than chasing lap times. Despite its vulnerable front spoiler and lack of stowage space, the 997 RS is a car that should be used and enjoyed for what it is.

The slightly softer GT3 makes a more practical tourer, but both the 3.6 and 3.8 RS are eminently usable and above all enjoyable as Marc Joly of *Flat 6* found. In 2022 he drove a 3.8 997 RS across southern France following mostly départementales and minor roads. He surprised himself not only at the fun he was able to have in a country where government has long been unfavourable to cars, but also at the positive reaction that his orange, be-winged Porsche received everywhere.

The stock advice is, as ever, buy well from a seller you trust, be pro-active about maintenance and ensure your lifestyle affords you time to drive the RS properly. Besides being at least depreciation-free, this – like all the 997 GT3 family – is a car whose talents are to be savoured. Like a retired competition greyhound, a 997 RS needs care and a diet rich in nourishment and exercise.

★★★★★

Ultimate 911 GT3 Collection

PORSCHE INDEX

Written by **Kieron Fennelly**

991.1 GT3 RS

A staggeringly responsive track special and a significant leap in technology compared with its 997 RS predecessor, here's all you need to know about the 991.1 GT3 RS

HISTORY & TECH

The RS moniker on Porsche dates from the 1950s, but the first Porsche RS that most think of is the celebrated Carrera 2.7 RS of 1972/3, built to qualify Porsche for Group 3 racing. Until the end of the air-cooled era, the RS remained the special race version, mildly tuned and lightened, and with reduced equipment and hardened suspension.

The advent of water-cooling in 1997 dictated a different approach. Production limitations and homologation requirements ruled out lightening, and the stiffer-chassis 996.1 GT3 on which the RS models would be based was 50kg heavier than the Carrera. The 996 GT3 RS of 2003, the inspiration of the Motorsport department's Andreas Preuninger, was an attempt to make a lighter, less-compromised GT3 track car and provide a homologation basis for the competition RSR.

The 997.1 GT3 RS presented in 2006 effectively dressed the GT3 in race gear, while the face-lift car of 2008 had more developed aerodynamic aids and used the 3.8-litre flat six that Porsche had been campaigning in the 996 RSR. It was some 25kg lighter than its GT3 equivalent.

After a hiatus of four years, the 991.1 GT3 RS represented quite an advance. Using the 991 Turbo body, the new RS was bigger than its predecessor. With more complex ducting around its rear flanks, a rear spoiler and vents in the front wings, the 991's downforce figure was twice that of the 997. Indeed, the 991 RS represented Porsche's most sophisticated aerodynamic package yet seen on a (nominally) road-going 911. Making extensive use of carbon fibre, in the weight-saving stakes the RS also went one better than its sister GT3 by having a magnesium rather than aluminium roof panel.

Even more impressive was the engine. Gone was the long-serving 'Mezger', superseded by a bespoke 4.0-litre flat six. With a 12.9:1 compression ratio, for homologation purposes it developed 500ps at 8,800rpm, though in fact nearer 515ps, revealed Preuninger. For a production engine this was an amazingly high-revving unit; however, after the crankshaft failures of the even-higher revving 3.8 of the 991.1 GT3 RS, Porsche was taking no chances. Besides bespoke titanium con rods and followers, and its own dry sump oil system, the crowning glory of the RS's 4.0-litre powerplant was the 919-derived crankshaft. This was an expensive component made from a special steel alloy, remelted several times to give lasting resilience.

The six-speed manual gearbox of the 911 RS finally gave way to a seven-speed PDK gearbox, while the suspension on the new 991 RS had bespoke, heavy-duty hubs and links in forged aluminium. Preuninger noted it was overkill for a road car, but it was necessary to homologate the suspension for the 991 RSR. In the cabin, buyers could choose between folding, fixed-height bucket seats, or leather and Alcantara buckets with electrical height adjustment from the 918 Spyder. ➲

THE VALUES STORY

First shown at Geneva in March 2015, the 991.1 GT3 RS was presented later than Porsche had planned because of the 991.1 GT3 recall. Once on the market, it was priced at £131,300 or €181,690. For comparison, a reasonably optioned 991.1 Carrera S then cost about £92,000.

Supply from Zuffenhausen (4,600 991.1 GT3 RSs would be manufactured by 2018) meant that waiting times were short by specialist car standards and prices didn't immediately take off. Conversely, when the 991.2 version arrived in 2019, neither were 991.1 values seriously affected and according to RPM Technik, a low-mileage, 991.1 GT3 RS with ceramic brakes and other extras would go for £180,000.

Since then, normal depreciation has reduced this slightly and prices in spring 2024 are around the £130,000-£160,000 range. Almost all offers involve cars with fewer than 20,000 miles.

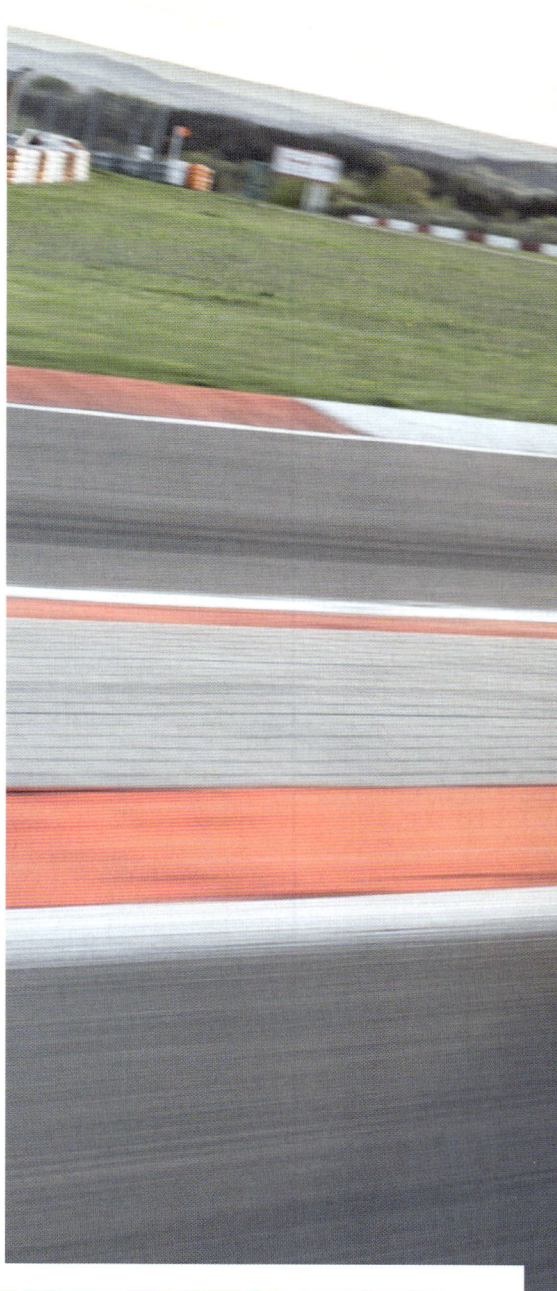

MARKET RIVALS

A mid-point of £160,000 yields a range of other track-oriented 911s

964 RS
The cheapest of the air-cooled RSs, not though a concours example at this price, the 964 provides a visceral ride and usable performance on a twisting road with all the 911 virtues of driver involvement. It'll also fit in the average UK garage.

997.1 GT3 RS
State of the art two decades ago, the almost entirely analogue RS 3.6 is a challenging drive, notably sharper than the plain GT3, but no quicker. A rare model (1,106 were built), RPM Technik has a top-money example at £110,000.

992 GT3
For the driver seeking the latest tech, the current 992 features such arcane road-going equipment as twin front struts. Offers begin around £180,000, but manual gearbox examples generally cost more. Still depreciating significantly.

996 GT3 RS
With only ABS and as analogue as the 964, this endearingly rough-edged "Preuninger special" has recently attracted collectors and values have increased. £185,000 should buy a pedigree example, but history will need careful investigation.

"It let you access another level of intensity of driving experience that no GT3 has ever approached"

DRIVING EXPERIENCE

It was immediately clear that in the years it had been absent from the market, the RS had become a much more advanced sports car. The advent of the larger, wider 991 chassis and longer wheelbase combined with a significantly more potent engine had enabled the engineers to boost 911 RS performance levels comprehensively.

In July 2015 Porsche introduced the new 911 RS to selected correspondents at Bilster Berg, the then-new Hermann Tilke-designed circuit near the German city of Paderborn. The scribes reacted unanimously: not only did the 991.1 RS look the part, it was the part. With nothing to do but point and squirt thanks to the ever-more effective PDK, acceleration was of a new order for a naturally aspirated engine. *Autocar* reached 60mph in 3.4s and 160mph in 23s, and commented that what the engine lacked in mid-range torque was more than compensated for by "sensational" throttle response and a power delivery at higher revs that was "hard-edged and spine tingling". Road roar from the 325/30/21 rear tyres was a small price to pay for the staggering adhesion they provided.

Driven fast round the challenging Bilster Berg, grip levels were "withering", thought veteran Porsche journalist Andrew Frankel, saying, "Traction is superb: this is a car with another level, a level beyond that of any 911 I have (ever) driven. It let you access another level of intensity of driving experience that no GT3 has ever approached." But he cautioned: "This is a car to drive extremely fast and if you are not going to drive extremely fast, there is little point in having one. It is a car that will challenge you… and it will not be shy about punishing mistakes." Frankel concluded, "There is a space at the top for one crazy car, one that takes you to the edge where you feel most alive: that car is the new GT3 RS."

BUYING A 991.1 GT3 RS

Although these cars are recent and many will have averaged fewer than 3,000 miles per year, purchasing still requires a close examination of the car and its history. The specialist dealers invariably have the better examples and the 65-plate RS on offer at RPM Technik is typical of the top end of the market. Priced at £155,000, this 6,700 mile RS has extended leather inserts and the very practical hydraulic front axle lift. This car's service record and condition should be impeccable. On the other hand, a 991.1 RS offered privately at, say, £125,000 may well offer much the same quality car as the dealer, albeit with higher mileage, but then it behoves the purchaser to follow good dealer practice and look at the chassis and underneath the carpets for creases or signs of welding.

Thanks to its superior, more mid-engine handling with higher corning limits, a 991 is less likely to have had track 'offs' than a 997 RS. Greig Daly of RPM Technik thinks that the 991's sheer speed means that few track-day drivers can get near its dauntingly high limit, and adds that the stability controls intervene more subtly than on the previous RS. Porsche development driver Lars Kern has told RPM that on the Nürburgring, he doesn't switch off the PSM; he can lap faster with the control on rather than off. For minor scrapes, bumpers and spoilers can be replaced, at worst affecting only originality. An RS that's had more fundamental workshop intervention is probably less desirable unless much cheaper. Signs of gravel spray, as Jason Shepherd of Paragon observes, give away a regular "circuit" car, but a GT3 RS that's seen significant circuit use isn't necessarily to be avoided. Porsche built them for the track in the first place – it simply needs to be reflected in the asking price.

Like the Mezger before it, even tuned to 120ps per litre, the 9A1 flat six is a robust unit. Porsche was confident enough to give the rebuilt 3.8 of the 991.1 GT3 a 10-year guarantee, and reports of maladies afflicting the 4.0 RS engine and PDK transmission are rare. RPM's experience is that even after seven years, its technicians aren't seeing any failures – even with cars driven regularly on the track. ➲

DESIRABLE OPTIONS

The choice of seat already mentioned is without additional cost. This model, like its sister 991.1 GT3, was never available with the manual gearbox; air-conditioning was a no-cost delete, but virtually all the RSs have it. The Clubsport cabin (bolt-in roll cage, battery master switch and driver's six-point harness) was as previously a no-cost option (and fills the rear compartment with scaffolding), but there were several payable options such as a leather rather than plastic dashboard finish that enabled owners to fine-tune the cabin. Standard chassis electronics included PASM and dynamic engine mounts, as well as traction controls. Furthermore, rear steer, as on the 991 GT3, was standard. The two most expensive options are the carbon ceramic brakes and the very practical front axle-lift which, like the carbon brakes, is also an aid to resale.

INVESTMENT POTENTIAL

When it was launched, the 991.1 GT3 RS offered remarkable value for money given its exotic technical specification and performance, and after the long gap since the 997 3.8 RS (excluding the collector-only RS 4.0) there was an eager market for a new GT3 RS. Unlike previous RS manufacture, production wasn't restricted to the 2,000 units Porsche had planned and ultimately 4,600 were built between 2015 and 2018 before the facelift 991.2 RS appeared. This meant market supply was maintained and there was no significant aftermarket for new cars quickly resold at a premium. No doubt this volume helped Porsche recoup the significant costs of the RS, but it also meant a more 'normal' depreciation curve.

Today's market offers aren't far from the original 2015 sales price and while a 991.1 GT3 RS purchased now would probably not lose any money, it would be misguided to think it might start appreciating like the fabled 4.0 997 GT3 RS. However, when a new 992 RS, which isn't much quicker, costs well over twice a £150,000 991.1 RS, the older car looks good value. 911

TOTAL 911 VERDICT

The question is whether for your, say, £150,000, the 991.1 GT3 RS is the Porsche for you. There are potentially stronger 911 investments: unlike the 997 RS 4.0, this GT3 RS is not a collectors' 911. As an everyday drive, the 991 RS might make sense for an owner living in wide, open spaces, but in much of southern Britain the Porsche's dimensions can make it a liability on narrow roads and in tight urban areas, where parking spaces seem to assume cars are all the same size as the original Mini. And in this context, as opposed to the paddock at Brand Hatch, the RS's aerodynamic accoutrements may appear ostentatious and could attract the wrong kind of attention.

Moreover, a car with such massive, eager performance is mostly going to be operating, at best, at one-third throttle. As other journalists have reflected, why have an extremely fast Porsche if you can't drive it extremely fast? So that leaves the track. A prospective owner needs to be honest about their abilities. If track outings are the main reason for purchase, would they be better off with a slightly more forgiving (and less expensive) mount like a GT3? But if our would-be purchaser can handle this RS, then the intensity of the experience is likely to be unmatched.

★★★★☆

ABOVE The 991.1 GT3 RS's bespoke engine can output 500ps, although Andreas Preuninger said this value was nearer 515ps

LEFT The 991.1 GT3 RS marked a big change for the RS recipe, using a Turbo-wide body and PDK gearbox

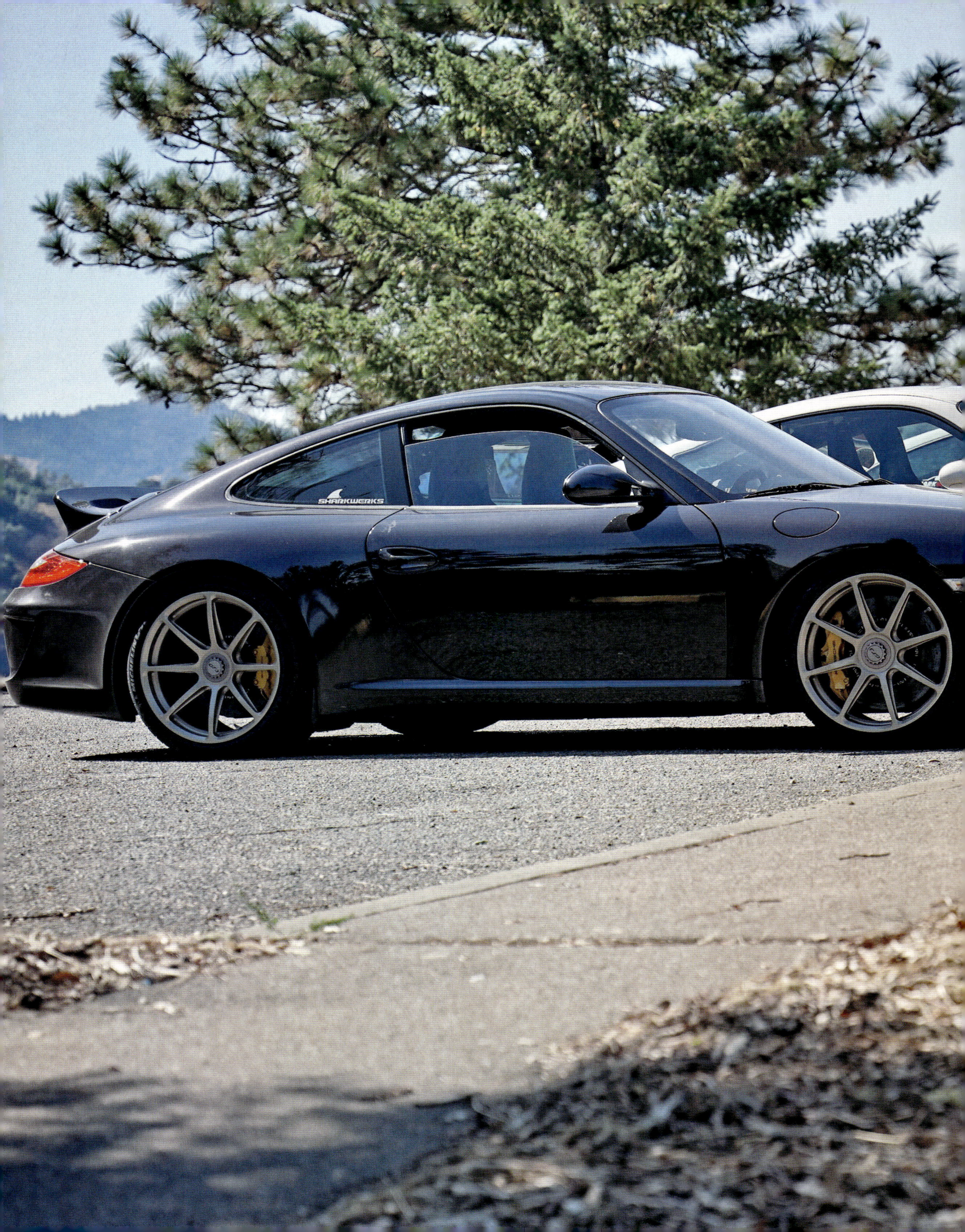
SHARKWERKS

MODIFIED & MOTORSPORT

110
GT3 HYBRID
Porsche used the GT3 platform to pioneer a hybrid racer back in 2010. Here's how it worked, but also why it failed

112
SHARKWERKS 997 GT3 TOURING
Porsche didn't make a GT3 Touring for the 997 generation, but US-based GT gurus, Sharkwerks, are implementing cutting-edge technology to bring it to enthusiasts

120
GT3 CUP CAR DRIVEN
We drive the ultimate iteration of GT3 – a Supercup-spec 992 – on the race track

997 GT3 R:
THE FIRST HYBRID 911

The 992.2 may be the first road-going hybrid 911, but the first ever hybrid 911 itself debuted some 14 years ago. Total 911 looks back on an intelligent idea stymied by self-serving regulations — Written by **Kieron Fennelly**

In the 2000s, to improve its image F1 turned to KERS. The kinetic energy recovery system harvested heat, otherwise wasted, from braking and transformed it to power, the driving wheels. Within a decade regenerative braking was a feature of mass-production hybrids that used batteries to store this energy and run electric-only for short periods.

The concept greatly interested Porsche and in 2009 Weissach bought a KERS rig from F1 pioneers Williams, with the idea of running a GT3 R so-fitted in competition. The system chosen by Weissach used a flywheel that was propelled by brake heat energy. Spinning between 28,000 and 36,000rpm, the flywheel, positioned in the passenger footwell, acted as a commutator supplying current to a pair of electric motors driving the front axle.

To function, the system relied on a complex combination of inputs from ABS, PSM, traction control, torque vectoring and the 4x4 transmission. The hybrid system boosted the 4.0-litre GT3 R's output from 480 to 620PS, and increased the GT3 R's weight by 150kg.

Unveiled publicly at the 2010 Geneva Show, the GT3 R Hybrid was described by Porsche as a racing laboratory to see what this technology could offer road-going sports cars. The Hybrid was blooded in a VLN four-hour race at the Nürburgring shortly afterwards. Its electric assistance incurred a 25kg ballast penalty, but the Hybrid finished third overall and again in the next VLN event. These

The First Hybrid 911 | 111

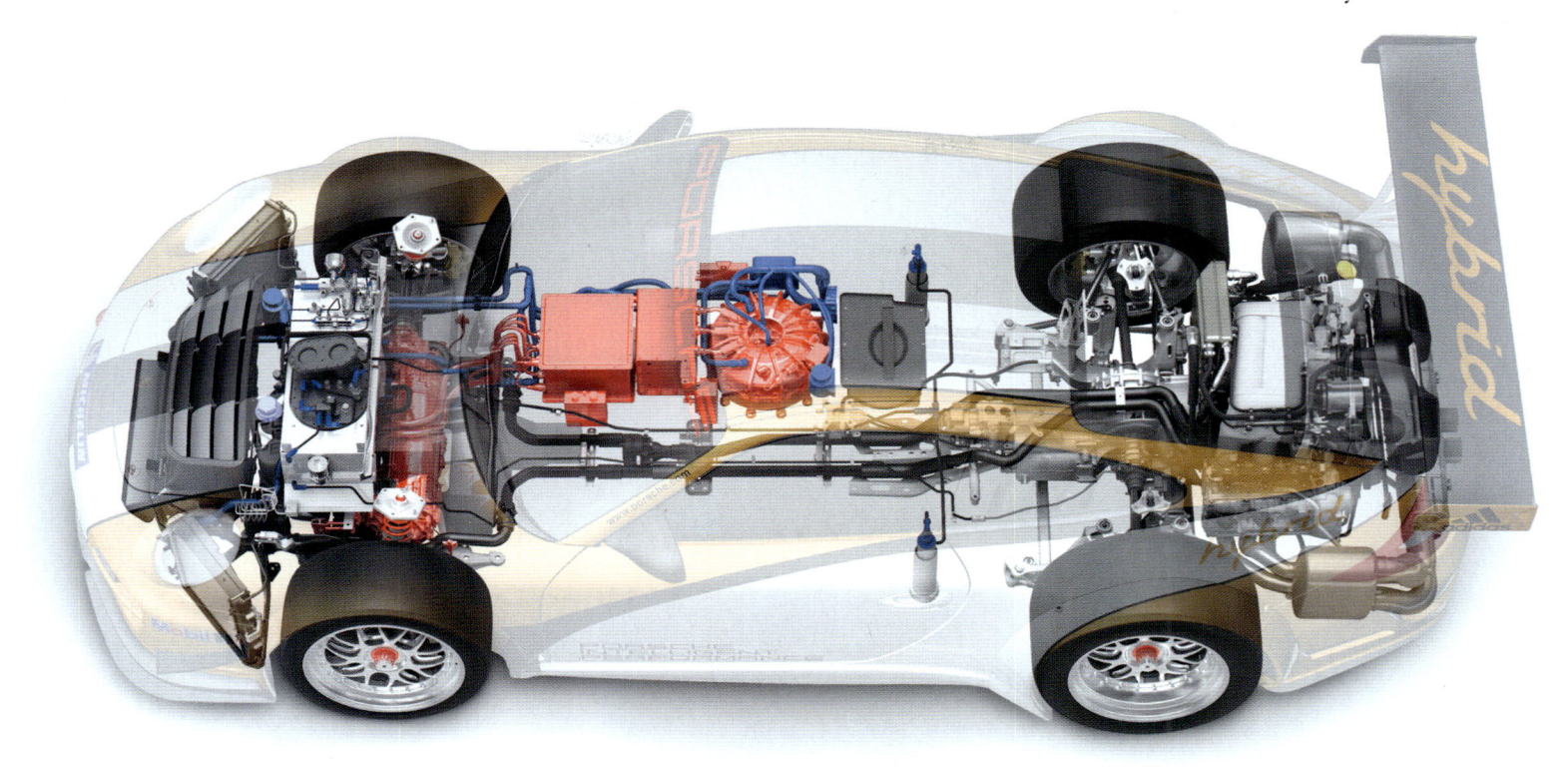

"The GT3 R Hybrid was described by Porsche as a racing laboratory to see what this technology could offer road-going sports cars"

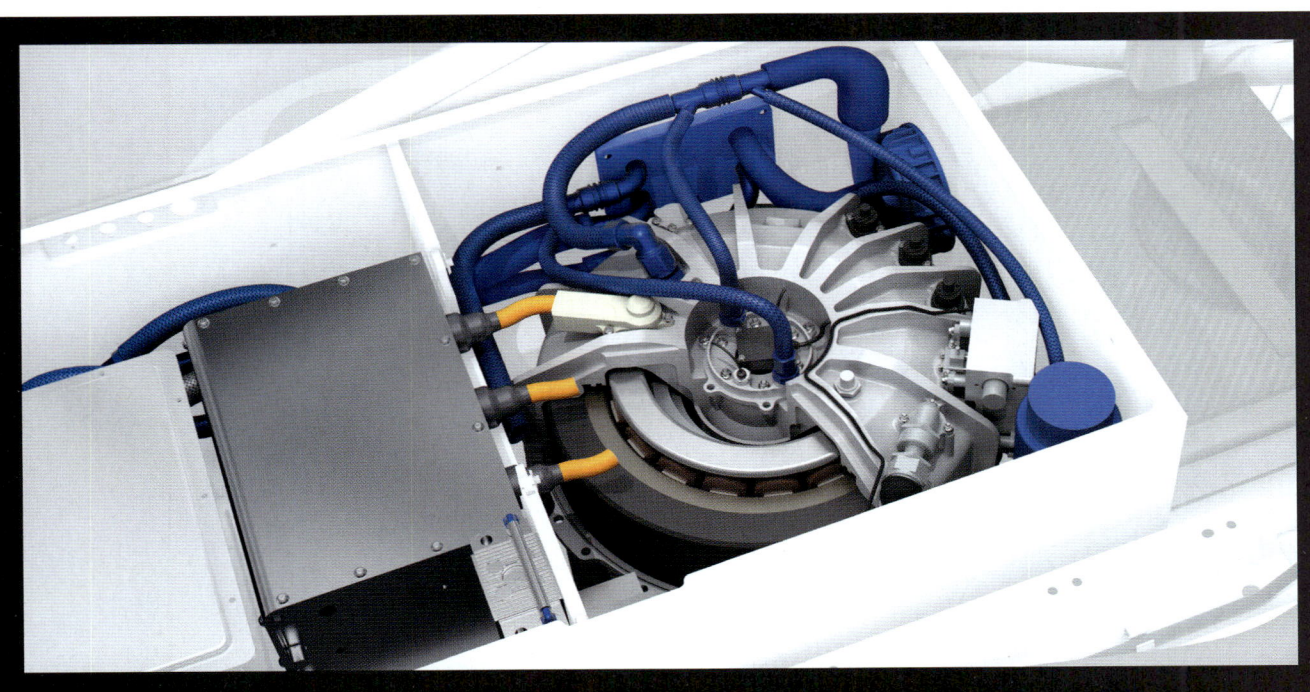

ABOVE The elements in red show how the electric system, powered by kinetic energy, were installed in a GT3 R

LEFT The hybrid system's flywheel span at speeds of between 28,000 and 36,000rpm, generating current to power a pair of electric motors that drove the front axle

'Ring 24 Hours, where nevertheless it was leading with two hours remaining before a valve spring broke.

During the summer the Hybrid was further developed with strengthened cooling systems and boost controlled by telematics. This meant the driver no longer had to remember exactly when to deploy the extra power. Finishes in the US and China in 1,000 mile and 1,000km races in October closed a season where Weissach felt it had proved a point: if no faster than rivals Audi or BMW, the Hybrid scored by requiring less fuel and fewer refuelling stops.

The problem was it didn't fit in any category. It was faster than the GTs, but as a 4x4 it was ineligible for this class. Weissach persisted, hoping the rules would evolve. Work over the winter to lighten the car saved around 140kg and although this incurred an intake restriction penalty costing 15PS, this was compensated by electric motors uprated by 80PS. Furthermore, the driver could now adjust the torque of each wheel individually.

Alas, the infamous balance of performance rules undermined the Hybrid once again. Its output was suddenly restricted just prior to a VLN race and narrowed again before the 'Ring 24 hours where, with no time for recalibration, the Hybrid had to be driven flat-out – losing much of its fuel economy advantage and twice breaking a transaxle. It competed once more, finishing 10th at Laguna Seca, before Porsche wound down development and quietly withdrew. Expensive to develop and, perplexingly, legislated out of its natural racing category, for some the Hybrid's fate recalled the 'exiling' of the 917 that took place 40 years earlier.

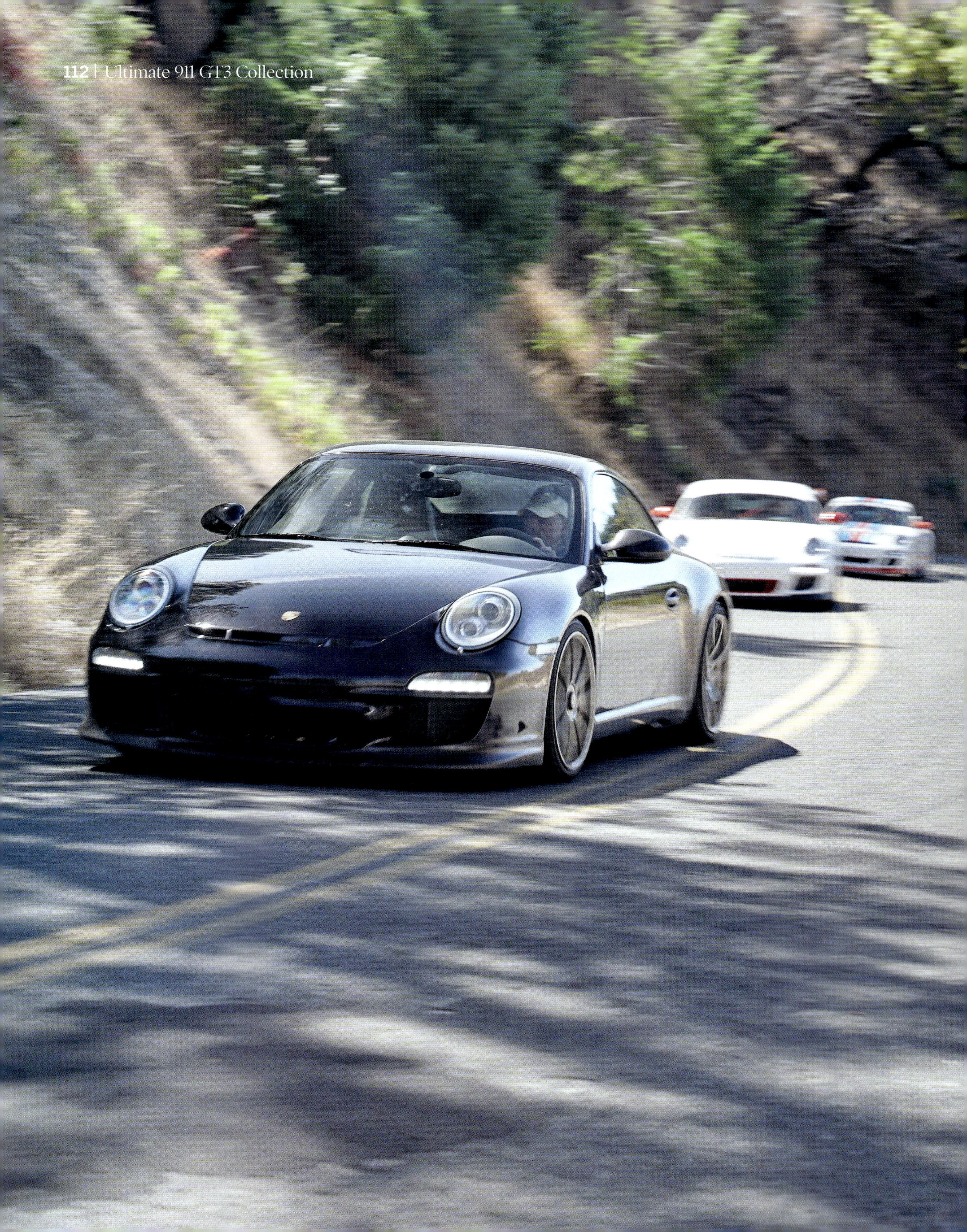

997 GOES TOURING

SharkWerks' new project will unlock exhilarating, streetable performance from Porsche's 997 GT3. Total 911 heads Stateside for an early drive

Written by **Lee Sibley** Photography by **Max Newman**

Gualala, northern California. Far away from the hubbub of The Golden State's major cities, and very far away from Stuttgart and the home of the sports car that this magazine is dedicated to, one could question what, indeed, has brought **Total 911** to this quiet, quaint pocket on Planet Earth? Allow me to explain.

Here, when travelling north, there's nothing but the Pacific Ocean to your left, tall Redwood trees to your right, and a ribbon-like strip of asphalt that wriggles on and on, dancing through the middle – not for mile after mile, but seemingly hour after hour.

Venture inland and the roads are even more twisty and enticing, of the sort that'll give you a solid workout of the arms, legs and brain, should you have the privilege of peddling a 911 along them at a half-decent pace. This is SharkWerks country: famous for a fine fettling of Porsche sports cars, from quarter-mile-conquering 996 Turbos to king-of-the-canyon-964s, SharkWerks has featured numerous times in these pages over the years. Needless to say, we really do rate their work highly here at **Total 911**.

You should already know, therefore, that the company's chef d'oeuvre is its 4.1-litre conversions of 997 GT cars. For good reason, too – to this day, a certain Riviera blue 997.2 GT3 RS 4.1 remains the most engaging, enthralling, positively intoxicating drive I've ever had in a 911, a good nine years after that drive took place.

So, when Alex Ross at SharkWerks announced he's personally acquired a 997.2 GT3 and is using it as the basis for a new project, it didn't take long for us to fire a message over, cap in hand, asking for an opportunity to get behind the wheel. And that's why snapper Max and I are in town, buckled into the GT3, and giggling like adolescents as we give ⮕

114 | Ultimate 911 GT3 Collection

chase to a trio of SharkWerks-tuned 997 GT cars swivelling through the bends ahead. They, like this GT3, have all been 'Sharkafied', yet while those GT3s are enjoying the mastery of being full-fat 4.1s, this one has only been lightly fettled at the time of writing, as SharkWerks begins its journey of touring and testing, before unleashing that special 4.1-litre magic.

The Porsche landscape has changed somewhat since SharkWerks began building out to 4.1-litres a full decade ago, with 997-generation GT3 RS base cars ever highly more sought after by collectors. That hasn't stopped Alex, James, Dan, Joan and the SharkWerks team, who have continued to build out factory 3.8 997 GT3 RSs and even 3.6-litre 997.1 GT3s to 4.1-litre specification, for a select few customers in Europe, Canada, Hong Kong and other US states. Porsche itself, meanwhile, has made 4.0 litres the customary displacement for its GT cars, also bringing to market a de-winged 'Touring' 991.2 GT3 in 2018, to widespread acclaim.

SharkWerks chases 'fun' rather than 'financial' and, with the 991 and 992 platforms not really hitting the spot for Alex and co, extending the 997 GT3 platform's breadth of capabilities in a similar way is the natural next step for a company intent on refining what it says is already a fantastic product. The goal here, then, is achieving high thrills between 30 and 90mph.

"It's a skunkworks project right now," Alex says over lunch a couple of hours later. "The plans for this car are to also outfit it with our 4.1 programme and short-gear stack. We are also exploring – though it's early days – a path similar to that of the factory 991.2 Touring.

"The ethos for this car is to make it a complete sleeper: removing the wing, a very subtle wheel change, as Forgeline GE1s look stock-like for example, and addressing the chassis to be more compliant and closer to the GTS, but with the Motorsport engine we love so much."

The idea is a sound one: modern 911 Q-cars have fascinated since the launch of the 2016 911 R, which paired a GT3 engine and manual gearbox (GT3s were PDK-only at the time) in a wide yet wingless body. To the everyman, it could well have been a Carrera.

The notion was expanded further by Porsche with the subsequent 991.2 GT3 Touring going wingless, while the 992-era has given enthusiasts another Touring (now with a choice of PDK or stick shift) and perhaps the ultimate Q-car: the S/T, which has an RS engine stuffed in the rear of its flat-backed body.

That car has rounded off a seven-year period of real thirst among enthusiasts for a strictly road-oriented Porsche product with Motorsport componentry. And why not? For some 911 fans, the idea of a race track has simply never appealed.

Rewind 15 years though, and Porsche's GT department offering wasn't as three-dimensional as it is now. The GT3 and RS were unapologetically crafted as track cars with licence plates: a euphoric experience on the circuit then, but on the street, they're a committed drive.

Alex and the SharkWerks team are well on their way to changing that. Even without that magical 543hp, 4.1-litre powerplant in place, this 997.2 GT3's current setup is enticing enough. Its powertrain has had some light tuning: a 997 GT3 4.0 RS clutch and single-mass flywheel has been deployed for sharper pick-up, and an uprated Guard LSD offers superior traction (and a little more longevity) on these tight, twisty roads. The GT3's soundtrack has been elevated too, with the company's own system giving a signature SharkWerks howl. Its chassis, meanwhile, features the company's customary RSS rear upper links and bump steer kit combined with Tractiv PASM-compatible coilovers, tuned with a DSC controller. It's a heady mix and, from behind the wheel, already feels like a 'GT3+', its playful charm more apparent at legal speeds.

Ahead of me, Ralph is leading the way in his white 4.1-litre 997.1 GT3. The tall uprights of its Cup wing sway left and right as the 911 pivots effortlessly through the bends that snake feverishly through a dense woodland. Behind, there's a Martini-

"The ethos for this car is to make it a complete sleeper… addressing the chassis to be more compliant and closer to the GTS, but with the Motorsport engine we love so much"

SharkWerks 997 GT3 Touring | 115

LEFT The winding roads and dense woodland of NorCal echo with the noise of a 997.2 GT3 chasing a 4.1-litre 997.1 GT3

BELOW Editor Lee Sibley surveys the 997.2 GT3 with SharkWerks' co-founder, Alex Ross

116 | Ultimate 911 GT3 Collection

RIGHT Big-winged 4.1-litre 997s contrast against the svelte ducktail on Alex's skunkworks 997.2 GT3

BELOW Ready for round two…? T911 heads out in the 997.2 GT3 for more fun on NorCal's sun-dappled roads

FACING PAGE A recently finished 4.1-litre conversion enjoys some running-in miles

SharkWerks 997 GT3 Touring | 117

"There's more than a hint of 997 RS 4.0 styling present, though somewhat outrageously, that's seen to be under-clubbing its performance"

livered 997.1 GT3, and a Carrara white 997.2 RS, which I can see clearly thanks to the removal of a fixed, factory wing on this 997.2 GT3. A Sport Classic-style ducktail sits in its place, its trailing edge visible from my interior mirror, now glinting in the bright sunlight as our road leaves the woods behind.

Who doesn't love a ducktail on a 911? It saves weight over the adjustable factory wing and, while it won't be doing the same job aerodynamically at higher speeds, at our targeted playground of 30-90mph the factory addenda would be superfluous.

So what's the result of this stage one fettling, as Alex puts it? It's certainly building on what is already a mighty setup from Porsche, and we can see why Alex and the SharkWerks team idolise the 997-generation so much. Its narrow body means road placement for a fast yet respectful line through the bends is easily found, with the era's hydraulically assisted steering gifting exemplary feel and weighting. Swiftly yet smoothly pushing the wheel through left and right turns, I'm reminded of just how accomplished this setup is. It really was a high watermark for Porsche 911 steering feel for many years (the subsequent 991-generation switched to electrically assisted steering, though it'd take until the following 992-generation for Porsche to truly perfect it).

Likewise, the six-speed manual shift is light years ahead of the messy seven-speed Porsche would introduce four years after this 997.2 GT3 was launched. The six's shifter has a lovely, crisp feel to each throw, matched by a clutch pedal that makes heel-and-toe a real pleasure to execute. The ratios themselves feel a little long on NorCal's darting blacktop, though that will, of course, be looked at once SharkWerks bring them closer together.

SharkWerks mods have only elevated the drive here, unlocking more of what we love about the 997 GT3. Throttle response, even at low rpms, is sharp (it'll be even more so with its incoming 4.1), and the top-end shriek emitted from the SharkWerks exhaust, in song with its three comrades, is best described as a religious experience.

Without a doubt though, this 997's genius lies in its chassis. The Tractiv suspension is brilliant: adjustable on the fly, there's a lovely suppleness to the ⟳

118 | Ultimate 911 GT3 Collection

BELOW Sunset over the Pacific closes out a day driving some fine examples of Neunelfers that SharkWerks has taken to the next level

ride at low and medium speeds, the sophistication of its execution easily putting it on a par with a modern 992 GT3 equivalent.

Staying between second, third and fourth gears and following the contours of our spaghetti-like road, the 997.2 is moving around beneath me in a talkative, playful manner, pivoting sharply when called upon, and giving chase to the superior metal we're holding court with. Alex says the car currently is too soft, but with one eye on bumpy UK roads back home, I think the setup could be perfect as-is.

A GT3 this may be, but right here and on these roads, I wouldn't care if I never saw a track again. Goodness, just how great will this thing be with a 4.1 out back?!

After one last stop in late afternoon, the opportunity to drive a full-fat GT3 RS 4.1 is presented. It takes but half a second to consider and accept. Being back behind the wheel of such an engineering marvel is not lost on me, the genius of which could never be replicated by Porsche on a mass scale, such are the development costs and time required to achieve it. It's also a decent yardstick as to what we can expect from Alex's 997.2 GT3 Touring project in due course.

The example I'm sat in has recently completed its SharkWerks makeover and, with the blessing of its kind owner, I'm helping to stick a few running-in miles on the odo. Finished in white, with grey and red decals, there's more than a hint of 997 RS 4.0 styling present, though somewhat outrageously, that's seen to be under-clubbing its performance. Not for the marginal gain in displacement, but the fact a SharkWerks 4.1 has anywhere between 80 and 110Nm more torque than the halo RS 4.0 across its entire rev range. A few miles on, and glorious memories of nine years prior flood back. Never mind that iconic collectible from Weissach, a peak 997 is shark-shaped!

The reality of SharkWerks' GT3 RS 4.1 creation is just as good as I'd remembered. Supremely powerful right through to its 8,800rpm redline, there's a scarcely believable, buttery-smooth execution to the way power is delivered. Free-revving, needle zipping around the tacho, its engine feels every bit as strong as the 543bhp Alex tells me it's kicking out, and there's plenty of fizz in the mid-range, which is often something of a dead zone for GT3 and RSs in stock form.

Zippy, zingy, and downright sensational, the 4.1-litre flat six is best characterised by the seemingly impossible contrast of the brute force of its power delivery against the delicacy of its throttle response. A beautiful harmony of balance and brawn, the SharkWerks 4.1 is fast art. The prospect of it appearing in the back of a road-focused 997.2 GT3 is genuinely tantalising – we can't wait to come back to NorCal to experience the fascinating, finished article. **911**

"A beautiful harmony of balance and brawn, the SharkWerks 4.1 is fast art"

Ultimate 911 GT3 Collection

GT3 Cup
Tested on Track

When you get an invitation from Porsche Supercup driver, Larry ten Voorde, to drive his 992 GT3 Cup, there's only one thing you can do: fasten your seat belt and enjoy the ride!

Written by **Sandor van Es** Photography by **GP Elite**

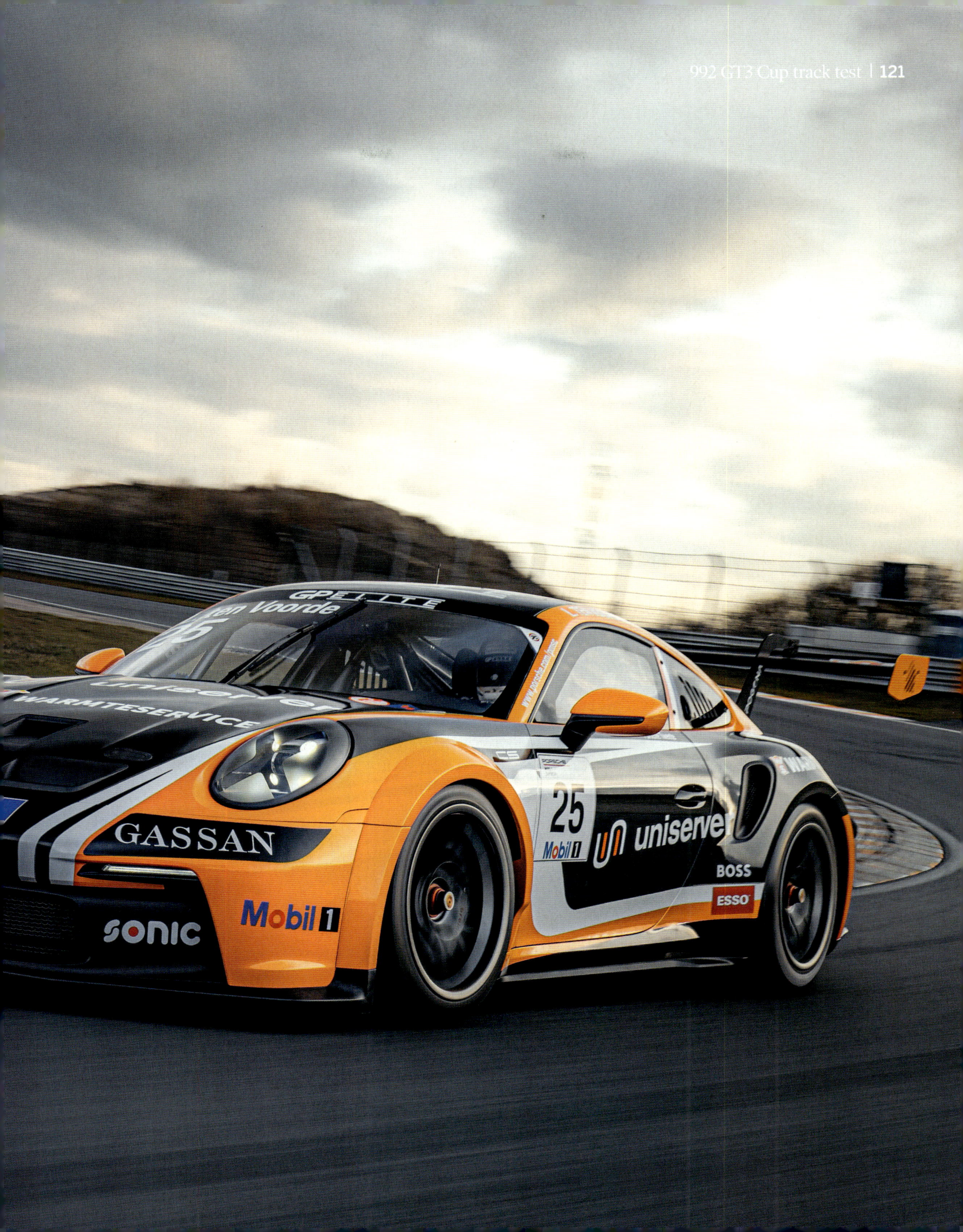
GASSAN
Mobil 1
sonic
25
uniserver
BOSS
ESSO
WARMTESERVICE

DRIVE TO PERFECTION
PORSCHE
GT3 Cup
VERWELIUS/BOUWEN
VERWELIUS/BOUWEN
CODE
0
Mobil 1
FRESH MINDS
25

992 GT3 Cup track test | 123

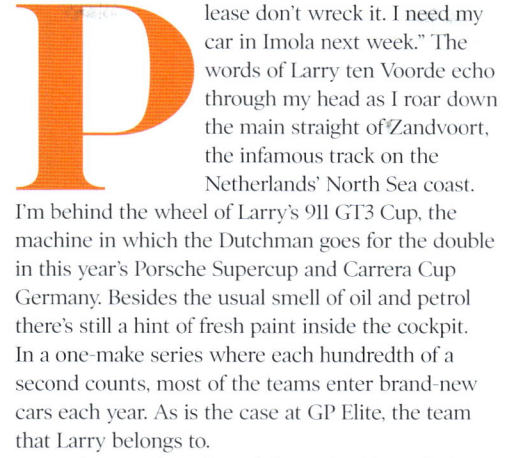

"Please don't wreck it. I need my car in Imola next week." The words of Larry ten Voorde echo through my head as I roar down the main straight of Zandvoort, the infamous track on the Netherlands' North Sea coast. I'm behind the wheel of Larry's 911 GT3 Cup, the machine in which the Dutchman goes for the double in this year's Porsche Supercup and Carrera Cup Germany. Besides the usual smell of oil and petrol there's still a hint of fresh paint inside the cockpit. In a one-make series where each hundredth of a second counts, most of the teams enter brand-new cars each year. As is the case at GP Elite, the team that Larry belongs to.

As the grandstands and the pit buildings flash by and I approach the Tarzan corner at well over 150mph, I realise it's quite brave of Larry to let me drive a car that got its shakedown just a few hours earlier. I don't want to give the mechanics extra work while dealing with a gruelling season in Supercup, so I hit the brakes at a safe 150 metres before Tarzan.

The brake bias is manually adjustable, and because of the cold asphalt conditions I've turned it slightly more to the rear. The unpowered installation needs quite a bit of muscle power from the right or left leg (depending on which foot you use). As long as the steering wheel is straight, they can be dosed perfectly. Nevertheless, you still need to carefully load up the front tyres – which have more camber than on a standard GT3 – before putting maximum pressure to the pedal, otherwise you risk a lockup. Once the car is settled, the deceleration is incredible. The stopping power of a regular 911 GT3 is already far above average, but on slicks it literally takes your breath away.

There's not a lot of time before the entry, so I quickly pull the left shift paddle four times. Wham-wham-wham-wham! With perfectly dosed automatic rev-blips, the transmission flicks down from six to two, lightning fast and without hesitation. In contrast to a PDK gearbox, this happens without the (automatic) intervention of a clutch. The GT3 Cup is equipped with a sequential dog-box. Controlled by an actuator, it changes gears directly – provided that the engine speed is properly synchronised in between. It's something you can leave to the state-of-the-art Bosch electronics of this racing car. The threatening, screeching sounds give an extra dimension to the violent sensations.

The brakes and gearbox are two important modifications compared to a regular GT3. In many other areas the Cup is quite similar. That's no surprise considering the fact that the road-going GT3 already has an overdose of race genes in its blood. Think of the double wishbone suspension with extra-wide track at the front, lighter body panels, enhanced cooling and a lot of extra downforce. With its dry-sump lubrication and several lighter and stronger rotating parts, the engine of the showroom version is track-ready as ⊃

"The deceleration is incredible... on slicks it literally takes your breath away"

well. Although alterations have been made to the air inlet and exhaust, the 4.0-litre engine still delivers 510hp and 470Nm. The only difference is that the six horizontal pistons rev up more easily. Thanks to the stripped interior, extra pieces of carbon fibre and the plastic side and rear windows, the weight of the GT3 Cup has been trimmed down to 2,778lb, which is – despite the roll cage – 287lb less than original. Add to this the racing chassis and the Michelin slicks, and you can imagine it hardly gets any better than this.

I enter the Hugenholtz, an oval-style 180 degrees left. The steering response is faster than any 911 I've driven before. I point the Porsche to the top of the corner to get the most out of the banking, using it to create even more grip. The moment I get pressed down in my seat and feel how the forces build up, I release the brakes and keep the momentum going. At three-quarters I force the car a few metres down to have a better line up for the exit. Thanks to the increased front end grip you can start to accelerate slightly earlier than before.

Nevertheless, it's still a 911, no matter how refined the seventh generation of this Cup racer is. To avoid mid-corner understeer you can't fully floor the pedal until the steering is almost straight. Once you start to press down your right foot you don't want to back off any more, because you want to take maximum advantage of the traction that's created by the weight of the engine at the rear tyres.

When you got the timing right, the car accelerates with brutal force. I'm catapulted up to the Hunserug dune. The engine joyfully screams as it shoots towards 8,400rpm. The LED lights at the top of the digital race display colour from green to yellow to red. I pull the right shift paddle and the transmission pops into the next gear. On top of the hill to fourth, in a blink of the eye the speed has increased to 100mph.

A lightning-fast section follows. Downhill to the flowing left-hander called Slotemaker, short shift to fifth and lift off the gas by 20 per cent to stabilise the car. Then uphill again, slightly right. Larry would take it all flat, but just before the top of the dune there's a bump at the racing line. "If you hit a kerbstone or bump at the right angle the car bounces only once and then settles, but if you hit it wrong you can get in trouble," he told me before. So I'd better be careful.

Even so, I fly over the top at 120mph. Brake briefly and not too hard, downshift once and then turn in at Scheivlak, an ultra-fast blind right-hander. The large rear wing does its job and presses the rear to the asphalt. This makes the timing of accelerating even more important. If you go on too early the nose will drift off to the outside and as a result you have to lift the throttle. Not a good idea at 95mph. No, better be a little patient and rotate the nose to the right direction before the right foot goes down again.

Braking briefly and downshifting once into Masters corner, then hit it hard and shift to second before the entry of Turn 9, sharp to the right. The key here is not to release the brakes completely during the turn-in and keep the front tyres loaded for optimum response and direction. Once more, timing and finesse are essential: if you brake too hard or too long, the rear will become loose. Luckily, these kind of dynamic balance changes are more predictable than in previous versions of the GT3 Cup, so you can play with your input to optimise the balance of the car through the corners.

Towards the Hans Ernst chicane, on the straight back to the paddocks, I briefly hit 125mph. Hard braking, downshift three times, sharp right-left. Patiently let the car roll through the second part of the chicane before flooring it. Again that acceleration, again that wonderful, atmospheric engine scream. Just before the Kumhobocht I briefly hit the rev limiter. Mid-corner, it's essential to keep the speed up, as you want to carry it with you to the main straight. Shift to fourth gear and after the Arie Luyendyk corner – since the renovation of 2020, spectacularly banked and easy flat – to fifth.

My GT3 Cup adventure at Zandvoort is coming to an end. Compared to the 991 Cup, the 992 is four seconds a lap faster, yet is better balanced and more accessible for inexperienced drivers. I'd love to fill up the tank and drive another session, but when I see the relief on Larry's face when I return to the pit lane, I realise I'd better not push my luck any further today. What a car, and what an experience! ➲

126 | Ultimate 911 GT3 Collection

TEN MINUTES WITH

Larry Ten Voorde

"My hard-headedness helped me to realise my dream"

On several occasions in the past, Larry ten Voorde had to sleep in his car because he couldn't afford a hotel. Nowadays the Dutchman, aged just 27, earns a living with racing in the Porsche Supercup and Carrera Cup Germany. He's a great example for youngsters who don't have wealthy parents to support them. We speak with the man himself to get the inside line on his career to date…

Total 911: You don't come from a wealthy family, so were your parents still able to help you at the start of your career?
Larry ten Voorde: I got some support during my karting years and also during my debut in Formula 4, in 2012. But then the money dried up quickly. I remember that after a crash at Spa we had to fix the car with duct tape and old parts from other teams. We also used to run discarded tyres from others. As you can imagine, I couldn't continue like that, despite being part of the talent pool of the Dutch motorsport association.

So in 2014 you had to quit…
I did, but I was already too addicted to motorsports to just let it go. I got in contact with GP Elite, which didn't run a racing team at that time, but instead just organised track days and racing courses. It asked if I wasn't shy of hard work and I responded that I would take anything. A few weeks later I was up to my ankles in the mud, assisting clients during a rally training.

It must have been hard to take a step back?
I just wanted to stay in this world, in whatever role it was. I owned an old Volkswagen Polo Wagon and travelled from one track to the other for GP Elite. Sometimes I slept in the back of my car because I had no money for a hotel. Building paddock tents, setting out cones, washing cars, cleaning the track, doing instruction work: I did everything they would ask of me.

How did the ball start to roll again?
As an instructor I was able to build up a lot of valuable relationships with all kinds of people. That was very helpful and motivated me even more to chase my dream. I asked my parents to allow me one year to restart my career. I took out a personal loan without considering the risks and without a Plan B. I raised just enough money to enter the German Porsche Sports Cup and from there things started to go again.

In 2017 you won the Rookie title in the Carrera Cup, in 2018/2019 you finished 3rd overall and in 2020 you were crowned both German and Supercup champion. The latter one was with GP Elite, which on top of the other activities had started a racing team…
From setting cones for them to winning the title – how cool is that?! In between the races I was still working there as an instructor, so our relationship got stronger and stronger. I really became part of the family.

Is this why you still race with GP Elite instead of perhaps pursuing a career with Porsche Motorsport?
I've had talks with them – quite serious ones. But I asked myself whether the total package was better than the one I have at GP Elite. After losing the Supercup and Carrera Cup titles last year, I decided to stay on and try to bring back both crowns to the Netherlands.

Do you have any advice for our younger readers who want to pursue a racing career as yours?
Be critical of yourself and avoid blaming others. Try a different approach if something doesn't work out. But most importantly, never give up!

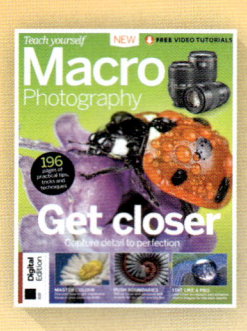

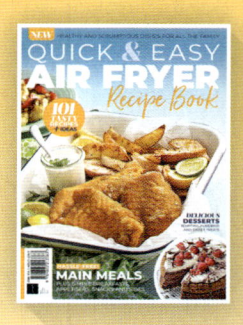

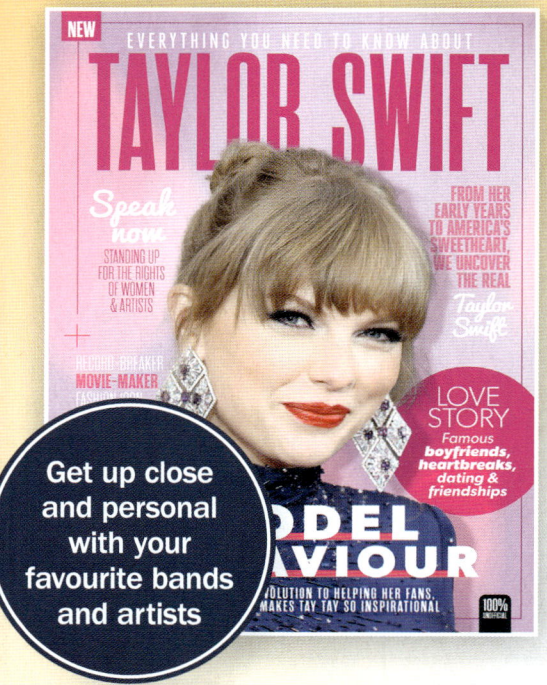

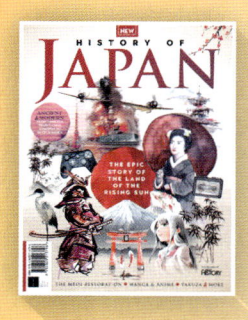

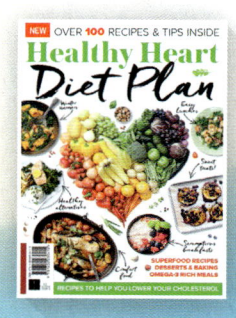

Discover helpful guides packed with tips and tricks to manage daily stress

Get up close and personal with your favourite bands and artists

Find inspiration and impress your dinner guests with recipes for every occasion

✓ Get great savings when you buy direct from us

✓ 1000s of great titles, many not available anywhere else

✓ World-wide delivery and super-safe ordering

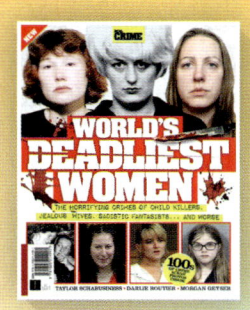

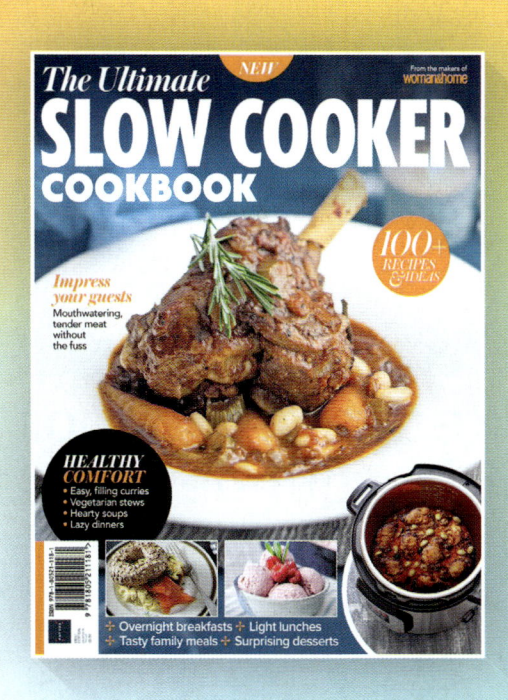

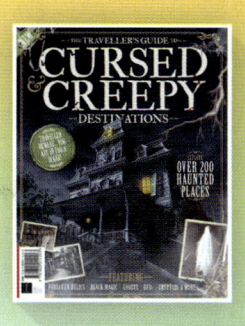

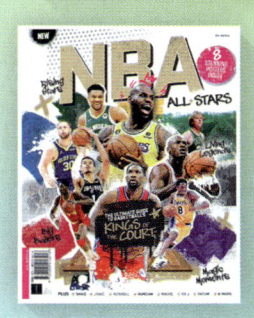

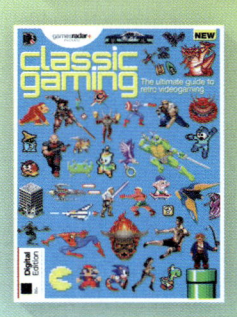

DISCOVER OUR GREAT BOOKAZINES

From history and music to gaming and cooking, pick up a book that will take your hobby to the next level

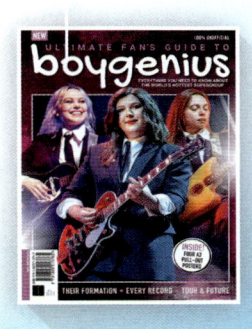

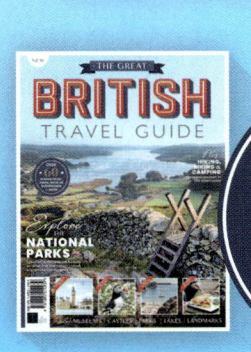

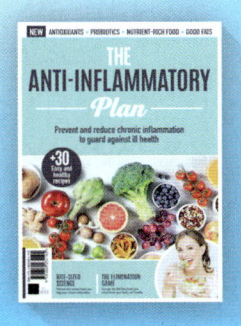

Discover fun facts and get creative with kids' activity books

Follow us on Instagram @futurebookazines

www.magazinesdirect.com
Magazines, back issues & bookazines.

Future

SUBSCRIBE & SAVE UP TO 61%

Delivered direct to your door or straight to your device

Choose from over 80 magazines and make great savings off the store price!

Binders, books and back issues also available

Simply visit www.magazinesdirect.com

✓ No hidden costs   🚚 Shipping included in all prices   🌐 We deliver to over 100 countries   🔒 Secure online payment

FUTURE

magazinesdirect.com
Official Magazine Subscription Store

ULTIMATE PORSCHE 911
GT3
COLLECTION

CELEBRATING 25 YEARS OF AN ICON

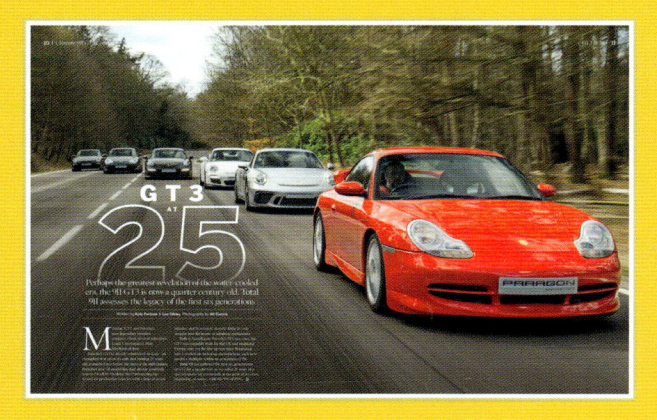

GROUP TESTS
Each generation driven back-to-back

BUYER'S GUIDES
Get yourself into the best example

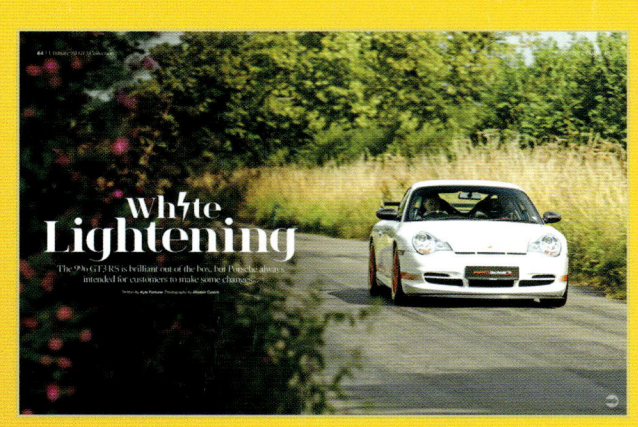

GT3 RS
Rennsport GT3s driven!

ROAD AND RACE
See the GT3 in action on road and track